Indigenous Peoples and Climate Change in Latin America and the Caribbean

Indigenous Peoples and Climate Change in Latin America and the Caribbean

Jakob Kronik and Dorte Verner

THE WORLD BANK
Washington, D.C.

GF
514
.K76
2010

ISBN: 978-0-8213-8237-0
eISBN: 978-0-8213-8381-0
DOI: 10.1596/978-0-8213-8237-0

Cover photos: Dorte Verner
Cover design: Naylor Design, Inc.

Library of Congress Cataloging-in-Publication Data has been requested.

Contents

Boxes

Tables

Foreword

As the Western Hemisphere slowly awakens to the nature of climate change, indigenous peoples have long been acutely aware of its impact. They observe directly its effects on the environment and the natural resources that sustain their livelihoods, while struggling to find appropriate responses to these new challenges.

This groundbreaking work, undertaken by the Sustainable Development department, delivers a first-hand account of how indigenous communities in Latin America and the Caribbean are affected by climate change and climate variability. The authors have written the first evidence-based book describing the social implications of climate change on indigenous communities in the region and have provided options for improving local resilience and adaptation policies. Across the region, indigenous peoples' perceptions of climate change correspond with observed changes in precipitation and temperature as measured by local weather stations and as described by global climate models. Despite the minimal levels of greenhouse gas emissions that result from traditional indigenous lifestyles, indigenous peoples often blame themselves for the climate change they observe around them. Motivated by this deep sense of relationship to their natural environment, they are looking for ways to modify their own behavior to help restore the balance between nature and humans.

This book bases its conclusions on new field research among indigenous communities in three major eco-geographical regions: the Amazon, the Andes and Sub-Andes, and the Caribbean and Mesoamerica. It finds major inter-regional differences in the impacts observed among areas prone to rapid and slow-onset natural hazards. In Mesoamerica and the Caribbean, increasingly severe storms and hurricanes damage infrastructure and property and cause the loss of productive land, reducing access to vital resources. In the Colombian Amazon, changes in the amount and seasonality of precipitation have direct and immediate effects on the livelihoods and health of indigenous peoples. Crops are failing more often, and the seasonal renewals of fish stocks are threatened by changes in the ebbs and flows of rivers. In the Andean region, water scarcity for crops and livestock, erosion of ecosystems, and changes in biodiversity threaten food security, both within indigenous communities and among populations who depend on indigenous agriculture, causing widespread migration to already crowded urban areas.

But the book reveals a common theme: indigenous communities find it difficult to adapt to environmental change in a manner that is consistent with their cultures and traditions. Not only is the viability of indigenous livelihoods threatened, resulting in food insecurity and poor health, but cultural integrity is also being challenged, eroding confidence in solutions provided by traditional institutions and authorities.

Cultural, social, and biophysical differences among indigenous peoples and their habitats shape how climate change affects their livelihoods. If climate change adaptation activities are to work for indigenous peoples, the initiatives need to be rooted in their social and ecological context. Adaptation strategies must be based on indigenous peoples' local knowledge about the environment in order to develop solutions that strengthen their resilience to climate change in a culturally, environmentally, and economically appropriate manner.

This book is an important contribution to our knowledge of climate change in general and a much-needed boost to our awareness of climate change's impact on indigenous peoples in particular. As such, it is also a guide for policy makers, indigenous peoples' organizations, and non-state actors working in the design of sustainable solutions to climate change adaptation.

Pamela Cox
Vice President
Latin America and Caribbean Region, World Bank

Acknowledgments

The study on which this book is based was written by Jakob Kronik and Dorte Verner. The book was developed and managed by Dorte Verner (task manager). The climate science section, which is included here as Appendix A, was prepared by Jens Hesselbjerg Christensen for a larger book, *Reducing Poverty, Protecting Livelihoods, and Building Assets in a Changing Climate: Social Implications of Climate Change in Latin America and the Caribbean.*

We would like to thank Inger Brisson, Sara Trab Nielsen, and Sanne Tikjøb for invaluable support in producing the original study. We acknowledge the indispensable research support of David Bradford, Juan Alvaro Echeverri, Jorge Cusicanqui Giles, Ricardo Hernández, and Evelyn Taucer Monrro. Special thanks to the team of advisers: Jocelyne Albert, Shelton Davis, Estanislao Gacitua-Mario, Andrew Norton, Walter Vergara, and Alonso Zarzar. We are grateful to peer reviewers Kirk Hamilton, Andrea Liverani, and Benjamin Orlove for their suggestions and comments. We also gratefully acknowledge helpful comments and suggestions from Maximilian Shen Ashwill, McDonald Benjamin, Carter Brandon, Isabelle Côté, Maninder Gill, Gillette Hall, Michel Kerf, Pilar Larreamendy, Mark Lundell, Augusta Molnar, John Nash, Frode Neergaard, Ben Orlove, Stefano Pagiola, Helle Munk Ravnborg, Tine

Rossing, Gustavo Santiel, and David N. Sislen. We are grateful to the management team of the Latin America and Caribbean Region of the World Bank, especially Pamela Cox, Laura Tuck, McDonald Benjamin, and Maninder. Gill for their support and encouragement. We would also like to thank Ramon Anria and Jorge Hunt for managing all the paperwork very effectively and Rachel Weaving for editing the study. We are also grateful for the support of Pat Katayama, Rick Ludwick, and Denise Bergeron of the World Bank Office of the Publisher, and Jeff Lecksell of the Bank's map services department. Finally, the World Bank is very grateful for financial support from Danida, Denmark.

Abbreviations

AdapCC	Adaptation for Smallholders to Climate Change
AMX	Amazon
AOGCMs	atmosphere-ocean global circulation models
CAM	Central America and Mexico
CDI	Comisión de Desarrollo Indígena
CEPAL	Comisión Economica para América Latina y el Caribe
CIDOB	Confederation of Indigenous Peoples of Bolivia
COP15	Conference of the Parties to UNFCCC (held in Copenhagen, Denmark, 2009)
CSUTCB	United Union Confederation of Campesino Workers of Bolivia
DFID	Department for International Development (United Kingdom)
ENSO	El Niño Southern Oscillation
EZLN	Ejército Zapatista de Liberación Nacional
FSLN	Frente Sandinista de Liberación Nacional
GCMs	general circulation models
GDP	gross domestic product
GHGs	greenhouse gases

GTZ	Deutsche Gesellschaft für Technische Zusammenarbeit (German technical cooperation agency)
IAS	Intra Americas Seas
IDEAM	Instituto de Hidrologia, Meteorologia y Estudios Ambientales
IGAC	Instituto Geográfico Agustín Codazzi
ILO	International Labour Organization
INEC	National Statistical Institute (Ecuador)
IPCC	Intergovernmental Panel on Climate Change
ISA	Instituto Socioambiental
ITCZ	Intertropical Convergence Zone
IWGIA	International Work Group for Indigenous Affairs
LAC	Latin America and the Caribbean
NARCCAP	North Amercian Regional Climate Change-Assessment
NGO	nongovernmental organization
NOAA	National Oceanic and Atmospheric Administration
PA network	Amazon Region Protected Areas Program
PRUDENCE	Prediction of Regional Scenarios and Uncertanties for Defining European Climate Change Risks and Effects
REDD	reducing emissions from deforestation and degradation
RAAN	North Atlantic Autonomous Region (Nicaragua)
RAAS	South Atlantic Autonomous Region (Nicaragua)
RCMs	regional climate models
SACZ	South Atlantic Convergence Zone
SAGARPA	Secretaría de Agricultura, ganaderia, desarrollo, rural, pesca y alimentatíon
SAMs	South American Monsoon System
SEGOB	Secretaría de Gobernación de Mexico
SINCHI	Instituto de Investigaciones de la Amazonía
SLF	Sustainable Livelihoods Framework
SRES	Special Report on Emissions Scenarios
SSA	Southern South America
SST	sea surface temperature
UCA	Universidad Centroamericana
UN	United Nations
UNAM	Universidad Nacional Autónoma de México
UNDP	United Nations Development Programme
UNEP	United Nations Environment Programme
UNFCCC	United Nations Framework Convention on Climate Change

CHAPTER 1

Introduction

Disorder in nature is a reflection of disorder in society.

—Nonuya man interviewed in the Colombian Amazon

Climate change is the defining development challenge of our time. More than a global environmental issue, climate change is also a threat to poverty reduction and economic growth, and it may unravel many of the development gains made in recent decades. Both now and over the long run, climate change and variability[1] threaten human and social development by altering customary means of livelihood and restricting the fulfillment of human potential (Verner 2010).

Indigenous peoples across Latin America and the Caribbean (LAC) already perceive and experience negative effects of climate change and variability. Although the overall economic impact of climate change on gross domestic product (GDP) is significant,[2] what is particularly problematic is that it falls disproportionately on the poor—including indigenous peoples, who constitute about 6.5 percent of the population in the region and are among its poorest and most vulnerable (Hall and Patrinos 2006).

This book examines the social implications of climate change and climatic variability for indigenous communities in LAC and the options for improving their resilience and adaptability to these phenomena. By social implications, we mean direct and indirect effects in the broad sense of the word *social*, including factors contributing to human well-being, health, livelihoods, human agency, social organization, and social justice. This book, much of which relies on new empirical research, addresses specifically the situation of indigenous communities because our research showed them to be among the most vulnerable to the effects of climate change.[3] A companion book (Verner 2010) provides information on the broader social dimensions of climate change in LAC and on policy options for addressing them.

Our hope is that by increasing understanding of the impacts of climate change and variability on indigenous communities, this book will help to place these impacts higher on the climate-change agenda and guide efforts to enhance indigenous peoples' rights and opportunities, whether by governments, indigenous peoples' organizations and their leaders, or nonstate representatives. Decisions made now about adaptation strategies, developing skills, and engaging with the broader community will determine the quality of life of the next generation. With more knowledge about how climate change affects indigenous peoples, governments and their development partners may be better able to understand and serve this group, providing them with the tools needed to break the downward spiral of poverty.

Indigenous Peoples of LAC

LAC's indigenous peoples are an important and diverse part of the population of many countries (table 1.1). Numbering about 40 million, the indigenous populations live in, and depend on, widely differing environments within the entire range of political and economic systems of the region.[4] The majority live in the colder and temperate high Andes (located between southern Colombia and northern Chile) and in Mesoamerica (between southern Mexico and Guatemala). Each of the more than 600 different ethno-linguistic groups in LAC has a distinct language and worldview, but most of the groups with distinct ethno-linguistic identities live in the warm tropical lowlands, especially in the Amazon rain forest. Although several indigenous cultures survive in urban settings, this book focuses on rural indigenous populations.

The livelihoods of many rural indigenous people are already fragile. Most of the region's indigenous people live in rural areas in extreme poverty, generally having little or no formal education, few productive resources, few work skills applicable in the market economy, and limited

Table 1.1   Facts on the Indigenous Peoples of LAC

Country	Number of peoples	Approximate number of indigenous individuals (1,000s)	Percent of total population
Argentina[a]	31	650–1,100	2.5–5
Belize[a, b]	2	55	17–20
Bolivia[c]	36	5,200	62*
Brazil[a]	225[f]	700	0.4
Chile[c]	8	1,060	6.5
Colombia[c]	92	1,400	3.4
Ecuador[c]	14	830[g]	6.8
El Salvador[a,b]	3	400	8
Guatemala[a]	20+[e]	6,000	60
Honduras[b,e]	7	75–500	1–7
Mexico[a,d]	61	12,400	13
Nicaragua[a]	7	300	—
Panama[a]	7	250	8.4
Paraguay[b]	20	89	1.7
Peru[b]	65	8,800	33
Suriname[a]	6	50	8
Uruguay[b]	0	0	0
Venezuela[a]	40	570	2.2
Total	644	38,800	n.a.

Sources: Authors' elaboration based on the following sources: (a) by self-identification International Work Group for Indigenous Affairs (IWGIA 2008); (b) Layton and Patrinos 2006; (c) national census self-identification; (d) Comisión de Desarrollo Indígena (CDI); (e) U.S. Central Intelligence Agency *World Fact Book*; (f) Instituto Socioambiental (ISA); (g) National Statistical Institute, Ecuador (INEC) 2001. Layton and Patrinos (2006) is the source for information not accompanied by a note.
Note: Though counting indigenous people is not an exact science, in this table the best available estimates have been gathered from various sources. — = Negligible. n.a. = Non applicable.
* Older than 15 years of age.

political voice. They depend strongly on natural resources for horticulture, fishing, hunting, and livestock herding, and a large percentage are agriculturalists, relying heavily on rain-fed maize (in Mesoamerica) and potatoes (in the Andes). Like many other poor people who depend on natural resources for their livelihood, their vulnerability stems from their dependence on fragile and threatened ecosystems as well as their comparatively limited access to infrastructure, services, and political representation.

But what sets indigenous peoples apart—and makes them especially vulnerable to climate change and variability—is the intimate ways in which they use and live off natural resources and their dependence on cultural cohesion. To maintain their livelihood strategies, they depend heavily on cultural, human, and social assets, including traditional knowledge systems and institutions, that are now under increased stress (Salick and Byg 2007).[5] Their knowledge systems are based on experiments with nature, juxtaposed with a stock of knowledge they have developed over time and passed on through generations. Their ability to predict and interpret natural phenomena, including weather conditions, has been vital for their survival and well-being and has also been instrumental in the development of their cultural practices, social structures, trust, and authority. The societal production of knowledge about nature's cycles has led to certain cultural practices. The practices, in turn, have resulted in the creation of cultural capital, which then is reproduced through practices and rituals. Cultural institutions are developed around these practices and rituals, serving to maintain, develop, and dispute information. These cultural institutions thereby contribute to the social generation of knowledge. All in all, their cultural institutions strongly affect indigenous peoples' natural resource management, health, and coping abilities.

Fundamental to many indigenous peoples' understanding of the relationship between society and nature is the notion that balances need to be maintained between the human, natural, and cosmological realms. These balances fluctuate continuously, and living and acting involves negotiating them, based on trusted social and cultural knowledge and practice. So when changes occur, for example, in climatic conditions, people look to themselves and their social institutions and practices to see whether aspects of the way they lead their lives are causing imbalance and need to be rectified. If this is not possible, they revert to other means to restore the balance. The cultural rituals and social corrective measures that are used vary from place to place, but they share the striving for balance between the social and the natural. To signal this understanding, this chapter opens with the quotation from a Nonuya man interviewed in the Colombian Amazon: "Disorder in nature is a reflection of disorder in society."

Climate Change and Climatic Variability in LAC

The impacts of climate change and variability are already being felt in the region. Initially, our research for this book was designed to gather indigenous peoples' perceptions and indications of climate-change impacts to come in the medium and long term. However, the large impacts that people said they were currently experiencing led us to examine which types of impact they perceive as most important to their livelihoods and institutions.

Comparisons show that indigenous people's perceptions of climate change and variability clearly correspond to meteorological data and to the projections in the *Special Report on Emissions Scenarios* (SRES) scenarios constructed by the Intergovernmental Panel on Climate Change (IPCC, 2001). Most of the indigenous people interviewed for this book related the increased variability and unpredictability of, for example, precipitation patterns to changes in long-term climate trends.[6]

Looking ahead, precise projections about climate change and variability within LAC cannot be made, because too little detailed historical information is available on the region's weather conditions, sea levels, and extreme events to allow robust regional climate models to be developed, and global climate studies yield relatively few robust statements and projections for the region (IPCC 2007a). Nonetheless, there is broad agreement about the trends: climate change is likely to have unprecedented economic, social, environmental, and political repercussions. Climate change "hot spots" in the region, and the threats they face, are discussed in box 1.1 and figure 1.1.[7]

Climate change and variability affect people's livelihoods, health, and general well-being in multiple ways. Figure 1.2 links greenhouse gas emissions resulting from human activity to environmental impacts, as presented in step 1. Environmental degradation affects, for example, water availability and quality for human consumption (including domestic, agricultural, and industrial use plus power generation) and terrestrial and marine flora and fauna ecosystems. This has social implications, as presented in step 2, affecting people's livelihoods, food security, and health. Excessive stress may cause additional impacts such as conflict, migration, greater income inequality, and poverty.

Five key climatic changes are occurring in LAC. Following the causal relations described in figure 1.2, each of the climatic changes and their environmental and social implications is outlined in more detail below:

- Rising atmospheric temperatures have major social impacts in LAC. They are already causing the melting of tropical glaciers, with implications for the amounts of water available for farming and livestock husbandry, domestic use, agriculture, power generation, and industrial use. Warmer air temperatures affect the geographical range of disease vectors such as malarial mosquitoes, with implications for human health. Warmer air temperatures also affect the range and yields of crops, with implications for the viability of traditionally grown crop varieties, and for agricultural practices, food production and trade, and

Box 1.1

Projected Climate Change and Climatic Variability in LAC

- The countries in the **Caribbean and around the Gulf of Mexico**, which are often assailed by intense hurricanes, can expect hurricanes to become even fiercer as a result of climate change. Other important issues are the destruction of coral reefs and mangroves and a growing threat to southeast Pacific fish stocks from rising sea-surface temperatures, leading to changed fish migration patterns. A rise in sea level would very likely bring flooding to low-lying regions such as the coasts of El Salvador and Guyana, further exacerbating social and political tensions in the region.

- For **Andean countries**, the most momentous climate-change impacts include major warming, changes in rainfall patterns, rapid tropical glacier retreat, and impacts on mountain wetlands. These factors, combined with increasing precipitation variability, will significantly affect water availability (IPCC 2007b) and may worsen the risks of migration and conflict.

- In the **Amazon region**, the most pressing issue is the risk that the rain forest will die back. Higher temperatures and decreases in soil moisture in the eastern Amazon region would lead to the replacement of tropical forest by savannah. If present deforestation trends continue, 30 percent of the Amazonian forest will have disappeared by 2050 (De la Torre, Fajnzzylber, and Nash 2009); its demise could trigger desertification over vast areas of LAC and even North America.

Source: Verner 2010.

food security. In adults, temporary malnutrition reduces body mass, immunity, and productivity, but the results are rarely permanent; in children, however, it can stunt growth, impede brain development, or cause death. Higher air temperatures also cause human health problems directly and can raise the mortality rate among infants, the elderly, and other vulnerable groups. Further, higher temperatures combined with decreases in soil moisture lead to deforestation, adversely affecting people's livelihoods in multiple ways.

- Rising sea-surface temperatures affect the viability and migration patterns of fish stocks and of coral reefs and mangroves, influencing the livelihoods of people who depend on their sustainable exploitation.

- Increases in the frequency or severity of natural hazards such as hurricanes, and changes in their geographical distribution, lead to higher death tolls and more damage to livelihoods, property, and production systems.

Figure 1.1   Climate Change Hot Spots in LAC

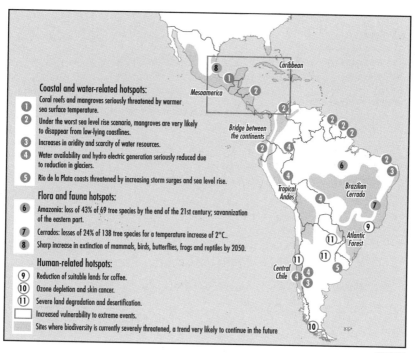

Coastal and water-related hotspots:

① Coral reefs and mangroves seriously threatened by warmer sea surface temperature.

② Under the worst sea level rise scenario, mangroves are very likely to disappear from low-lying coastlines.

③ Increases in aridity and scarcity of water resources.

④ Water availability and hydro electric generation seriously reduced due to reduction in glaciers.

⑤ Rio de la Plata coasts threatened by increasing storm surges and sea level rise.

Flora and fauna hotspots:

⑥ Amazonia: loss of 43% of 69 tree species by the end of the 21st century; savannization of the eastern part.

⑦ Cerrados: losses of 24% of 138 tree species for a temperature increase of 2°C.

⑧ Sharp increase in extinction of mammals, birds, butterflies, frogs and reptiles by 2050.

Human-related hotspots:

⑨ Reduction of suitable lands for coffee.

⑩ Ozone depletion and skin cancer.

⑪ Severe land degradation and desertification.

☐ Increased vulnerability to extreme events.

Sites where biodiversity is currently severely threatened, a trend very likely to continue in the future

IBRD 37799
MAY 2010

Source: Adapted from IPCC 2007.

Weather-related disasters affect human health not only through encouraging the transmission of diseases, but also through higher incidences of food insecurity caused by the erosion of crucial environmental and physical assets. Changes in the predictability of seasonal weather patterns have big implications for both commercial and subsistence agriculture, sometimes rendering traditional routines obsolete and wiping out crops, with implications for food prices, nutrition, and food security.

- Changes in precipitation amounts and patterns lead to more droughts—affecting rural livelihoods and food security—and more floods—affecting livelihoods, property, production systems, and food security for both rural and urban populations. Already, the widening incidence of drought has been a key reason why one-third of the region's rural youth have migrated to towns and cities over the past 20 years. Both droughts and floods also augment the risk of water- and vector-borne diseases.

- Rising sea levels lead to more floods and storm surges, affecting livelihoods, property, and production systems, as well as the habitability of

Figure 1.2   Nexus between Climate Change and Social Implications

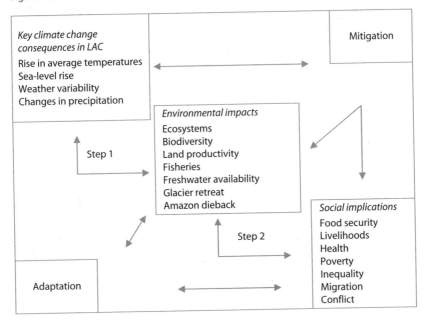

Source: Verner 2010.

coastal settlements; more beach erosion affecting the habitability of coastal settlements; salinization of soil, limiting or eliminating its use for agriculture; and the intrusion of saltwater into aquifers relied on for drinking water.

Determinants of Vulnerability

Clearly the social impact of climate change and natural hazards depends not just on biophysical factors but also on the resilience of people and their institutions to shocks. This book views vulnerability to climate change and variability as susceptibility to harm from exposure, sensitivity, and lack of capacity to adapt to environmental change (Adger 2006).

To permit a systematic analysis of vulnerability to the effects of climate change, we use a slightly adapted version of the United Kingdom's Department for International Development's (DFID's) Sustainable Livelihoods Framework (DFID 2001). This is a tool for assessing the vulnerability of different socioeconomic groups and their capacity to cope in the presence of shocks and to adapt to changing trends—which is very

Figure 1.3   The Sustainable Livelihoods Framework

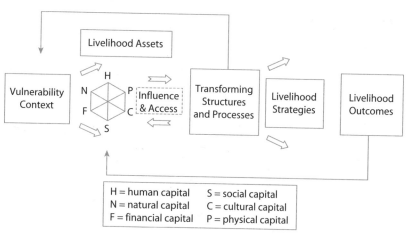

Source: Augmented from DFID 2001.

important from a climate-change perspective (figure 1.3). For this book, we have added cultural capital (Bourdieu 1973, 1986; Bourdieu, de Saint Martin, and Clough 1996) to the five livelihood assets considered in DFID's original framework. Our field research revealed that, particularly for indigenous people, the cultural dimension of livelihood strategies and social institutions is crucial for understanding the impacts of climate change and climatic variability. A more in-depth description of each of the livelihood assets is provided below.

- **Vulnerability context** refers to the environment in which people live. People's livelihoods and the wider availability of assets are affected by all types of external trends and shocks, including seasonality and climatic variability.
- **Livelihood assets** of individuals, households, or communities include physical, financial, human, social, natural, and cultural capital. The more assets a person has available, the less vulnerable he or she is. To maintain a livelihood successfully a person needs some combination of the different types of assets, because these complement and can to some extent substitute for one another. Access to livelihood assets determines a person's level of resilience and adaptive capacity with respect to climate change. The reverse relationship is just as important; climate change and variability can affect access to assets.

- **Transformational structures and processes** at play within the community are the institutions, organizations, policies, and legislation that shape livelihoods. They operate at all levels—from the household, community, and municipality to the national and international—and in all spheres—from the most private to the most public. They effectively determine access (to various types of assets, to livelihood strategies, and to decision-making bodies and sources of influence); the terms of exchange between different types of assets; and the returns (economic and otherwise) on any given livelihood strategy.
- **Livelihood strategies:** The Sustainable Livelihoods Framework (SLF) provides insight into how vulnerability, assets, structures, and processes influence livelihood strategies and how these strategies may be improved.
- **Livelihood outcomes:** The SLF examines how livelihood strategies, given the other factors, result in different livelihood outcomes and how the livelihood outcomes feed back into available assets, creating either a virtuous or vicious circle.

Risk or vulnerability factors increase the likelihood that a person or community will suffer negative outcomes. Experience shows that the risks associated with climate change increase when combined with poverty, poor governance, and poorly maintained infrastructure. Vulnerability to these risks often depends strongly on local social, political, and economic realities and government policies.

Protective factors increase the likelihood that a person or community will transition successfully. Important protective factors against the negative impacts of climate change exist at the household, community, and societal level. Protective factors include good public policies, such as provision of public health services, education, or social protection schemes; social connectedness, whether to relatives, neighbors, civil society organizations, or government agencies; solid and well-maintained infrastructure; good governance; and healthy public finances. Over generations, indigenous communities have developed and actively maintained sets of regularized practices, amounting to cultural institutions, which shape their management of natural resources and help them adapt to social and other changes. The active maintenance and respect for these institutions is a key feature of the well-being of such communities.

Climate change compounds people's existing vulnerabilities by eroding livelihood assets. For the poor and for indigenous people in particular, the detrimental effects of climate change on the environment erode a broad set of livelihood assets—natural, physical, financial, human, social, and cultural.

The livelihoods of many indigenous communities are already fragile, being based directly on natural resources in environmentally fragile areas. And many indigenous communities already face threats to their traditional livelihoods and indeed their territories from factors such as the encroachment of agriculture and mining activities into forest lands, political unrest, illegal coca cultivation, and pressure from the military and police.

Approach and Outline of Study

To select cases for detailed analysis we began by distinguishing the main types of climatological impacts that affect LAC.[8] On this basis, Latin America and the Caribbean are subdivided into the corresponding major ecogeographical regions: the Amazon, the Andes and sub-Andes, and the Caribbean and Mesoamerica.[9] In each of these regions, areas and peoples of particular interest were selected for further analysis. The aim was to uncover whether indigenous peoples are especially vulnerable socially to the effects of climate change and variability, and if this is the case, then why, how, and under which circumstances. We undertook qualitative analyses of relationships in LAC countries between climate change and the socioeconomic situation of indigenous peoples. We also did extensive field surveys in indigenous communities across the region (notably in Bolivia, Colombia, Mexico, Nicaragua, and Peru) to investigate how individuals and communities perceive and respond to climate change and variability. For details of the fieldwork, see appendix B.

Most of the changes and types of impact discussed in this book have effects across many population segments and environmental and social conditions. Therefore, the information gathering and data analysis sought to be responsive to factors such as gender, age, and the social and ecological situation of indigenous peoples, including land availability, degree of contact, and cultural resourcefulness. An effort was made to understand indigenous people's perceptions and perspectives from their worldviews. But because of the diversity of peoples and circumstances, a book like this can offer only a partial perspective on the wide range of realities that indigenous peoples experience—and the ways they interpret climatic changes and the impacts these impose.

The volume is in six chapters. Chapters 2, 3, and 4 are regional case studies of the social impact of climate change on indigenous peoples of the Amazon, the Andean and sub-Andean region, and the Caribbean and the Central American isthmus, respectively. Chapter 5 offers a comparative analysis across the three regions and includes a note on mitigation in an indigenous-peoples context along with a debate on Reducing Emissions from Deforestation and Degradation under the United Nations (UN) Framework Convention on Climate Change. Chapter 6 concludes, offering operational recommendations and suggestions for future research.

Notes

1. Climate change refers to long-term change, that is, slow changes in e.g. mean annual temperatures and precipitation levels leading to sea-level rise, melting of glaciers, etc. Change in climatic variability refers to increased unpredictability with respect to seasons, rainfall, hurricanes, etc., and is largely related to the El Niño Southern Oscillation (ENSO).

2. For LAC as a whole, cost estimates of damage from climate change and climatic variability vary from 1.3 to 7 percent of GDP by 2050 if no adaptation takes place. For 2000–05, the annual average cost of climate-related damage may be 0.7–0.8 percent of GDP (Nagy and other 2006).

3. This book addresses climate-change mitigation and any potential associated negative social impacts only to a limited degree, despite the obvious importance of such impacts. It does not address in detail issues of energy, forests, or infrastructure because these are the subjects of other ongoing work.

4. There is much debate about the numbers of indigenous individuals in Latin America. National statistics estimate around 28 million, while other estimates range from 34 to 60 million, reflecting the use of various definitions and calculation methods (Layton and Patrinos 2006). The public policy ramifications of estimating population size are widely recognized as influencing the degree to which access to services, resources, and rights are given or limited. International Labour Organization (ILO) Convention 169 promotes the use of self-identification, while other bodies favor the dominance of language or parents' language as the determining factor. For further reading, see ILO Convention (No. 169) on Indigenous and Tribal Peoples in Independent Countries.

5. IPCC's *Fourth Assessment Report* (2007) points out that there are "sharp differences across regions and those in the weakest economic position are often the most vulnerable to climate change and are frequently the most susceptible to climate-related damages, especially when they face multiple stresses. There is increasing evidence of greater vulnerability of specific groups..." IPCC (2007) makes specific mention of indigenous peoples and traditional ways of living only in the cases of polar regions and small island states.

6. This is similar to observations made by Thomas and others (2007) in South Africa.

7. Our assumptions on climate change in the region are detailed in appendix A, which summarizes existing knowledge about recent climate change and variability and the most likely future climate change and variability in the LAC

region, building on the recent IPCC *Fourth Assessment Report* (2007) and recent literature.

8. Hurricanes and intensified storms, increasing temperatures with corresponding glacier retreat, increased drought, and unpredictable variations in precipitation regimes.

9. Mexico and Central America.

References

Adger, Neil. 2006. "Vulnerability." *Global Environmental Change* 16: 268–281.

Bourdieu, Pierre. 1973. "Cultural Reproduction and Social Reproduction." In *Knowledge, Education and Cultural Change*, ed. Richard Brown. London: Willmer Brothers Limited.

Bourdieu, Pierre. 1986. "The Forms of Capital." In *Handbook of Theory and Research for the Sociology of Education*, ed. J. G. Richardson, 241–258. New York: Greenwood Press.

Bourdieu, Pierre, Monique de Saint Martin, and Laurette C. Clough. 1996. *The State Nobility*. Palo Alto, CA: Stanford University Press.

CDI (Comisión de Desarrollo Indígena). 2008. Mexico City, Mexico.

De la Torre, Augusto, Pablo Fajnzzylber, and John Nash. 2009. "Low Carbon High Growth: Latin American Responses to Climate Change." Overview to the report *Low Carbon High Growth: Latin American Responses to Climate Change*, World Bank Latin American and Caribbean Studies. Washington, DC: World Bank.

Department for International Development (U.K.). (2001). "Sustainable Livelihoods Guidance Sheet." http://www.nssd.net/pdf/sectiont.pdf.

Hall, Gillette, and Harry Anthony Patrinos, eds. 2006. *Indigenous Peoples, Poverty, and Human Development in Latin America*. New York: Palgrave.

INEC (Instituto Nacional de Estadística). 2001. VI Population Census. Quito, Ecuador.

ISA (Instituto Socioambiental), Brazil. Personal communication.

IPCC (Intergovernmental Panel on Climate Change). 2001. *Special Report on Emission Scenarios*. Geneva: IPCC.

————. 2007a. *Climate Change 2007: The Physical Science Basis*. Contribution of Working Group I to the *Fourth Assessment Report* of the IPCC. Geneva: IPCC. http://www.ipcc.ch.

————. 2007b. *Climate Change 2007: Impacts, Adaptation, and Vulnerability*. Contribution of Working Group II to the *Fourth Assessment Report* of the IPCC. Geneva: IPCC. http://www.ipcc.ch.

————. 2007c. *Synthesis Report: An Assessment of the Intergovernmental Panel on Climate Change*. Geneva: IPCC. http://www.ipcc.ch.

————. 2007d. "Forestry." In *IPCC Fourth Assessment Report, Working Group III*. Geneva: IPCC. http://www.ipcc.ch.

IWGIA (International Work Group for Indigenous Affairs). 2008. *The Indigenous World Yearbook*. Copenhagen: IWGIA.

Layton, Heather Marie, and H. A. Patrinos. 2006. "Estimating the Number of Indigenous People in Latin America." In *Indigenous Peoples, Poverty, and Human Development in Latin America*, eds. Gillette Hall and H. A. Patrinos. New York: Palgrave.

Nagy, M. Aparicio, P. Barrenechea, M. Bidegain, R. M. Caffera, J. C. Giménez, E. Lentini, G. Magrin, A. Murgida, C. Nobre, A. Ponce, M. I. Travasso, A. Villamizar, and M. Wehbe. 2006. *Understanding the Potential Impact of Climate Change and Variability in Latin America and the Caribbean*. Report prepared for the Stern Review on the Economics of Climate Change. http://www.hm-treasury.gov.uk/media/6/7/Nagy.pdf.

Salick, Jan, and Anja Byg. 2007. *Indigenous Peoples and Climate Change*. Oxford, U.K.: Tyndall Centre for Climate Change Research.

Thomas, David S. G., Chasca Twyman, Henny Osbahr, and Bruce Hewitson. 2007. "Adaptation to Climate Change and Variability: Farmer Responses to Intra-seasonal Precipitation Trends in South Africa." *Climate Change* 83: 301–322.

U.S. Central Intelligence Agency. *World Fact Book*. https://www.cia.gov/redirects/factbookredirect.html.

Verner, Dorte. 2010. *Reducing Poverty, Protecting Livelihoods and Building Assets in a Changing Climate: Social Implications of Climate Change in Latin America and the Caribbean*. Washington, DC: World Bank.

CHAPTER 2

Indigenous Peoples of the Amazon

The elders say that the Father of Friaje has been killed.

—Ticuna man from the Amazon River

Severe climatic changes are occurring in the Amazon basin, as documented by numerous climate models (IPCC 2001, 2007). Various phenomena are in play, including the El Niño Southern Oscillation (ENSO), La Niña, winds from the Antarctic, and warming from the Caribbean. The Colombian part of the Amazon is the focus of this chapter, as this area encompasses a wide range of the current and potential types of climate change and variability phenomena; types of indigenous peoples; and types of socioeconomic, cultural, and environmental conditions in the LAC region. Generally, very little information is available on how indigenous peoples in the Colombian Amazon perceive and react to climate change and variability. Therefore, this chapter rests on case studies of a few selected groups, supplemented with relevant information available on other indigenous groups. Examples from other Amazonian countries are brought into the analysis where relevant.

The Colombian Amazon covers an area of about 400,000 km², or around 6 percent of the 7 million km² of the Amazon, located in the northwestern section of the Amazon basin. This is a region of high natural and cultural diversity. Though humidity and precipitation are higher there, and forest conservation is much better than in most of Amazonia, the tropical forests of the Colombian Amazon are already profoundly impacted by climatic variation caused by the ENSO. These effects are exacerbated by deforestation and forest fragmentation, the extent of which varies depending on natural, cultural, and social variations within the region. At present the advance of deforestation is clustered in the western Andean foothills and is primarily the result of extensive cattle ranching and illegal coca plantations. Two different precipitation regimes characterize the northern and southern part of the Colombian Amazon.

The diverse indigenous population of the Colombian Amazon numbers around 100,000, consisting of 52 ethnic groups from 13 linguistic stocks and 10 isolated languages (table 2.1).[1] The northern precipitation regime is home to most of the indigenous peoples of the Colombian Amazon and the neighboring parts of the Brazilian and Venezuelan Amazon; the southern regime encompasses the southernmost indigenous peoples of Colombia. There is also a clear contrast between indigenous peoples in the western part of the Colombian Amazon, where colonization, armed conflict, illegal crops, and deforestation have had a great impact, and indigenous peoples in the eastern part, where the populations maintain greater territorial and cultural autonomy.

Combined, these local differences mean that there is great variation in indigenous people's control over, access to, and dependence upon natural, social, cultural, human, and economic resources. This variation is likely to

Table 2.1   Facts on the Indigenous Peoples of Amazon Countries

Country	Number of peoples	Approximate number of indigenous individuals (1,000)	Percent of total population
Bolivia[a]	36	5,200	62.0[*]
Brazil[b]	225[d]	700	0.4
Colombia[a]	92[e]	1,400	3.4
Ecuador[a]	14	830[f]	6.8
Paraguay[d]	20	89	1.7
Peru[c]	65	8,800	33.0
Suriname[c]	6	50	8.0
Venezuela[b]	40	570	2.2

Sources: Own elaboration based on the following sources: (a) national census self-identification; (b) by self-identification (IWGIA 2008); (c) Layton and Patrinos 2006; (d) Instituto Socioambiental (ISA); (e) indigenous peoples organizations; and (f) INEC 2001.
Note: (*) Older than 15 years of age.

condition the impacts experienced and foreseen from climatic change, as well as indigenous peoples' capacity to engage in effective adaptation. While the above-mentioned variation will contribute insights into the effect of climate change on various livelihood strategies, depending on different combinations of resources, conditions, and opportunities, it is not likely that all variation is captured.

The first section of this chapter describes indigenous people's perceptions and interpretations of changes in temperature and seasonality, comparing these with meteorological data. The second section analyzes the direct effects of these changes on livelihoods. Section 3 describes the broader social context affecting indigenous peoples in the region, including the advance of colonization and deforestation, and section 4 reviews the institutions—both indigenous and in mainstream society—that affect their livelihoods. The fifth section documents indigenous peoples' current adaptation and survival strategies to cope with these changes, highlighting adaptations of horticultural systems. The concluding section emphasizes that the effects of global climate change and variability in the Amazon region should be assessed together with those of other social changes affecting the region.

Threats and Perceptions of Climate Change

Temperature

Key informants interviewed from several areas of the Colombian Amazon note that temperatures in Amazonia have an annual cycle of variation, with lower temperatures in the middle of the year (winter in the austral hemisphere), but they agree that since the year 2000, the weather has in general become noticeably hotter.[2] As stated by one interviewee from the Caquetá River basin, one indication is that previously, the heat of the day would gradually decline from 6 p.m. until 9–10 p.m., after which point the night would be cool. Nowadays, the heat does not dissipate even after midnight.

These perceptions correlate well with local climatological data. Figure 2.1 shows the historical monthly temperatures for Leticia (on the Amazon River) and Puerto Leguizamo (on the Putumayo River) as compared with average monthly temperatures for the more recent period 2000–2007. Seasonal temperatures behave differently in the two river basins, which belong to two different seasonal regimes, as explained below; both regimes see declines in the middle months, but in the Amazon River area the highest temperatures are in October, while the area by the Putumayo River is warmest in January. In the Amazon River area, average

Figure 2.1   Monthly Temperatures on the Amazon and Putumayo Rivers (°C)

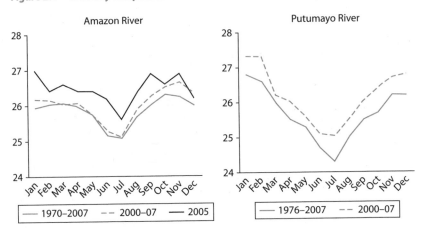

Source: Colombia, Instituto de Hidrologia, Meteorologia y Estudios Ambientales (IDEAM) 2008.
Note: Left graph (Amazon River): Meteorological data from the Aeropuerto Vasquez Cobo (Leticia) station (04°12′ S, 69°57′W), median monthly temperatures. Right graph (Putumayo River): Meteorological data from the Puerto Leguizamo station (00°19′ S, 74°46′W), median monthly temperatures.

annual temperature in 2000–2007 was higher during the final months of the year compared with 1970–2007, and 2005 (a drought year) saw an increase of almost 1°C throughout the year. In the Putumayo River area, the average annual temperature was about 0.5°C higher in 2000–2007 than in 1976–2007.

Seasonality

Different seasonal regimes exist in the north and south of the Colombian Amazon—the dividing line is located south of the equator, along the 2°S parallel (figure 2.2). Figure 2.3 shows historical data for precipitation and river levels for both regimes. The differentiated regimes mean that climate changes and variations are not uniform for all areas; for example, the 2005 drought that caused extensive fires in the Peruvian, Brazilian, and southern Colombian Amazon was not felt as strongly in the northern Colombian Amazon.[3]

Livelihoods in the Colombian Amazon are affected more by changes in precipitation and seasonality than by the impact of increasing temperature (Bunyard 2008; Salick and Byg, 2007). Indigenous people perceive that traditional seasonal variations in climate are now amplified.

The annual succession of seasons is of utmost importance for indigenous peoples. This rhythm orders the timing of the horticultural cycle and the ritual practices that are believed to help prevent illnesses and promote human well-being, and it is crucial for the reproduction of

Figure 2.2   Colombian Amazonia: Northern and Southern Seasonal Regimes

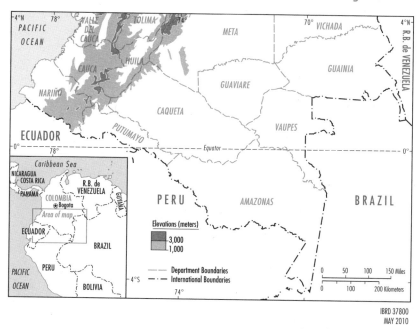

IBRD 37800
MAY 2010

Source: World Bank.

wildlife. Most indigenous peoples in the Amazon basin see the annual rhythm as having been established by their Father Creator since the beginning of time. Indigenous peoples are acutely aware of the ecological and ethical underpinnings of the annual cycle and of the adverse effects of its alteration. Based on observations by the Nonuya and other peoples who live along the Caquetá River (roughly on the equatorial line), box 2.1 summarizes the annual ecological calendar as it has repeated itself for generations.

Traditionally the annual ecological cycle has been affected by recurring interannual variations. Every few years, for instance, exceptionally high floods (called *conejeras* in Spanish) occur, and it is not unusual for the seasons to start earlier or later than their expected time. Indigenous peoples are aware of astronomical features such as the Pleiades and the Milky Way (Orlove, Chiang, and Cane 2002), even though skies are generally cloudy in the Amazon, and these provide constant markers against which they can detect the fluctuations of the seasons.

Indigenous peoples have observed that recent years have seen ecological markers occurring abnormally early or late, decoupled from the weather or season they used to mark, and different in timing, kind, or

Figure 2.3   Colombian Amazonia: Northern and Southern Seasonal Regimes: River Levels and Precipitation

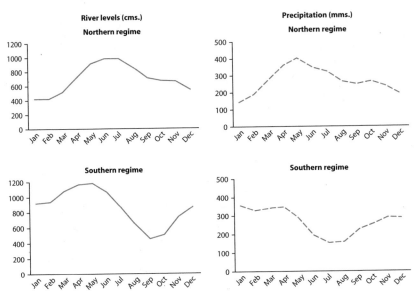

Source: Colombia, IDEAM 2008.
Note: Northern regime: Meteorological data from Araracuara Station (0°37′S, 72°24′W), mean values for 1979–2007. Southern regime: Meteorological data from Aeropuerto Vasquez Cobo (Leticia) station (0.4°12′S, 69°57 W), mean precipitation values for 1970–2007; and Yahuarcaca station (04°11′S, 69°57′W) mean river levels for 2000–2007.

Box 2.1

The Annual Ecological Calendar (as It Should Be)

The beginning of the indigenous peoples' year is the time of cold weather, called *friaje* in Spanish (*friagem* in Portuguese, *aru,* in the Vaupes region), in the month of July. Normally, *friaje* takes place two or three times, with the last episode being the strongest and coinciding with the ripening of the fruits of the miriti palm, *Mauritia flexuosa,* which grows in extensive patches over muddy soils. The miriti fruits are an important source of food for animals and birds. The *friaje* lasts for three or four days; it comes with winds and drizzle and with a remarkable decline in temperature (about 16°C). Fish die in lakes and rivers, and birds do not sing. For indigenous peoples, the *friaje* is the time of "menstruation" of the earth. The winds of this short but powerful time are conceived of as a man who comes to impregnate the earth and to fecundate everything—wild fruits, terrestrial fauna, fish, and humans. The winds

(continued)

Box 2.1 *(continued)*

also bring life and protection to humans, who can gain strength and renewed skills by bathing early in the cold weather. The winds of *friaje* arrive when the rivers are at their highest level and mark the end of the rainy season.

Friaje is followed by a succession of small dry seasons, which are named after the fruits that ripen in each—summer of *Pouteria caimito*, summer of *Ananas comosus*, and so forth—separated by rainfall. These brief dry seasons extend from August through September and are the time for slashing and burning plots that have lain fallow, in preparation for planting. This is followed, in October and November, by the "Time of Worms," when the forest fills with worms that feed on the wild fruits. It is a time of thunderstorms, followed by intense heat, and it favors the appearance of viral diseases. The river rises and recedes. As the river dries up, then comes the summer of *Poraqueiba sericea* (a wild and cultivated fruit), in December; the great summer of the peach palm *Bactris gasipaes*, in January, when the river is at its lowest level; and the summer of *Macoubea witotorum*, later in January. This is the dry season, when the main gardens, slashed in mature forest, are burnt, aided by the heat and the strong winds that blow constantly from the east. Just as the winds of *friaje* are called the Impregnator Father, the winds of the great summer are thought of as a Mother who comes to prepare the new gardens.

In March, the great rainy season commences, when new gardens are planted and the river begins its steady rise. It rains through April and May, when the river floods the lower and higher alluvial plateaus and inundates extensive areas of the forest. During this time, most wild fruits, fed by the water, ripen and fall into the rising river. They attract fish that come to the surface and disperse all along the flooded forest floor to feed on the plentiful food and to lay their eggs. The rise of the river stagnates with the arrival of the time of *friaje*, and a new cycle begins. The river waters again begin to recede, flushing away the contamination of the world.

Source: Narrated by Hernan and Eladio Moreno, Nonuya Indians from the Caquetá River basin.

intensity than the normal recurring interannual variations. According to their observations, the following changes are occurring:

- *Summer (dry season):* There is no longer a clearly distinguishable season with strong winds; there is heat, but not coupled with winds. Successions of dry days do take place, but outside their usual time of the year, which is July–August in the southern regime and January–February in the northern regime. In 2007, summer did not arrive at all, in either the northern or the southern regions.

- *Cold season:* For indigenous peoples, a decisive marker of the vitality of the whole ecological system is the appearance of the cold winds and drizzle of *friaje* (cold season). The winds are thought to blow away illnesses from the forest, expel the contamination of nature, and fertilize and impregnate flora and fauna (box 2.1). The *friaje* seems to occur earlier than usual and is less strong or is very short. If the winds of *friaje* do not blow strongly enough, they do not "purify" the trees, whose fruits therefore cannot ripen. As one Ticuna man from the Amazon River put it, "The elders say that the Father of *Friaje* has been killed."

- *Precipitation:* The rainy season is in disarray: there are no distinct dry or rainy seasons. "It rains when it should not, it is hot when it should not be," one man from Araracuara told us. Though the time of our interview (April 2008) was supposed to be the peak of the rainy season in the Igaraparana River basin (northern regime), three days of blue skies and intense heat were followed by storms and then days of menacing cloud cover that resulted in merely drizzle. At the same time, the Amazon River area (southern regime), which supposedly was in a dry season, was experiencing very heavy rains (figure 2.4).

Figure 2.4   Caquetá River (Araracuara) Monthly Precipitation (mm)

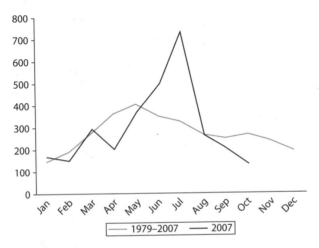

Source: IDEAM, Colombia 2008.
Note: Meteorological data from Araracuara station (0°37'S 72°24'W), monthly median precipitation for years 1979–2007 and 2007.

- *River levels:* This prominent seasonal marker has shown all sorts of alterations, with differences between the northern and southern regimes. In the northern region, in 2005, the river flooded and fish hatched as normal, but then the river suddenly receded before the fish were sufficiently mature, thus killing them (figure 2.5).

- *Drought and fires:* 2005 was an exceptionally dry year in southern and western Amazonia. Indigenous peoples along the Colombian Amazon River were affected by smoke coming from extensive fires that occurred in southwestern Brazil and eastern Bolivia and reached up to the Amazon River.[4] As shown in figure 2.6, this drought was not felt in the regions under the northern regime, but it did prevent the river from flooding that year. In Puerto Leguizamo (a) and Leticia (c), the precipitation regimes were inverted. At the Puerto Arica station (b), located exactly on the dividing line between the two regimes, the drought was still strongly felt.

Indigenous Interpretation of Climate Change

Society, not nature, is at the heart of Amazonian indigenous peoples' interpretation of the current situation—weather included. Indigenous peoples of the Amazon tend to attribute ongoing climatic alteration to human responsibility and inappropriate behavior. Though they have heard, in the

Figure 2.5   Caquetá River Monthly Levels (cm)

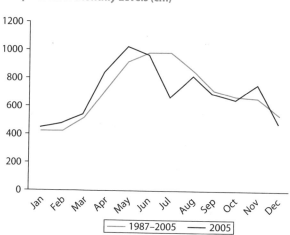

Source: IDEAM, Colombia 2008.
Note: Meteorological data from Las Mercedes station (0°32′S 72°10′W), river levels for years 1987–2005 and 2005.

be synchronized with precipitation, which helps the ripening of wild fruits on which fish feed in flooded areas. In 2007, a high flood in the Caquetá and other northern rivers allowed fish to reproduce success-fully after they had failed to do so the previous five years. On the Amazon River (southern regime), flooding has not been sufficient since 1999, and this has directly affected fish reproduction. The abnormal behavior of river levels has also directly affected the reproduction of turtles, which require a timely decrease in the river level for beaches to appear for them to lay their eggs on (box 2.2).

An early, short, or weak *friaje* coupled with an erratic succession of dry and rainy seasons directly affects the harvests of wild fruits.[5] Failed fruit-tree crops are observed among many indigenous peoples in the Amazonian and also the sub-Andean region, from the Aymara, Afro-Bolivian, and Yungas of Bolivia, to the Achuar and Shuar of Ecuador and Peru, to the

---

Box 2.2

The Effects of Seasonal Change on the Giant Turtle, an Endangered Species

The timely descent of water levels is required for the reproduction of turtles, par-ticularly of the giant turtle, *Podocnemis expansa,* an endangered species. As the river recedes and the beaches appear, turtles place a first clutch of eggs on the beaches in November (northern regime); normally, this first clutch is washed away by a final rise of the waters. Then, the turtles place a second clutch of eggs, which is successful, as the river finally dries up and the dry season sets in. In 2005, in the Caquetá River, the second clutch of eggs was also washed away by an abnormal rise of the river.

The giant turtle has been totally or nearly extinct from most white water Colombian and Brazilian Amazon rivers as a result of overexploitation. The Nonuya and Miraña Indians have been working with Colombian environmental authori-ties in conservation programs for this species in the Caquetá River, where the species still exists. They watch the beaches where the turtles regularly place their eggs and prevent traders or strangers from plundering the nests. They have also attempted to breed the turtles in artificial reservoirs. It is possible to salvage turtle nests from flash floods by moving them to higher ground. A Colombian non-governmental organization (NGO) carried out initiatives of this nature in the 1980s and early 1990s, but these experiments, which were criticized by Indians, did not continue.

Nonuya, Witoto, Muinane, and others of Colombia. The failure affects animals that feed on fruits and people who depend on both animals and fruit for food. Two prominent crops, which mark the peak of the rainy season and coldest month *(Mauritia flexuosa,* the wild miriti palm) and the peak of the dry season and hottest month *(Bactris gasipaes,* the cultivated peach palm) have shown abnormal behavior recently: in 2007, the peach palm harvest was very poor, and in 2008 the harvest of the miriti palm was unexpectedly early (in February instead of July).

Effects on Horticulture

Increasing heat affects indigenous people's horticultural work. Garden plots are open areas where heat is strongly felt. At midday it has now typically become too hot to stand barefoot on the garden soil. People are now forced to return early from their gardens, thus shortening or even eliminating the hours they can spend working the plots.

Higher temperatures, combined with changes in precipitation and seasonality, directly limit the early growth and success of plants. Some crops need to be replanted two or three times. Manioc, the staple crop in Indian gardens, is more resistant to heat and drought and grows well even in poor soils. Though manioc guarantees a source of carbohydrates, the increased heat threatens crop and nutritional diversity.

Indians work at least three types of horticultural plots:

- Slashed plots in mature forest on *terra firme* soils (acidic clay soils with poor nutrient levels). These require a longer drying period and need to be burnt to improve pH and enrich the soil. These plots support the most diverse Indian gardens.
- Gardens on alluvial plateaus on quaternary soils enriched by periodic floods. These gardens may not require burning, but they support only crops that can be harvested before the plateau floods again. These are the most productive but least diverse gardens.
- Slashed plots in secondary forest. Although these are less labor intensive than plots in mature forest and do not require a prolonged dry season to burn successfully, slashed plots in secondary forest are less productive and more prone to weed invasion. They are short-lived and less diverse than gardens in mature forest.

All three gardening systems have been directly affected by ongoing changes in seasonality and temperature.[6] Garden plots in mature forest depend on the regular occurrence of a predictable season of dry, hot days with constant winds, which guarantee a complete burnout of the plots. The unpredictability of such a season has caused slashed plots to be burnt incompletely or not at all. In the former case, Indians have to cut and stack wood to build fires and complete the burning process, greatly adding to the work needed. This type of garden is mostly worked by indigenous peoples of the northeastern Colombian Amazon, who have access to plenty of mature forest; indigenous peoples closer to urban or colonization areas no longer work this type of horticultural plot, as their access to high forest is severely restricted. Diversity-rich gardens in mature forest are important not only as a source of food but also for ritual and ceremonial life. Thus climate change and variability directly affect both cultivated diversity and ceremonial vitality in the more traditional areas.

The more productive but less diverse gardens on alluvial soils are threatened by unpredictable changes in river levels. Flash or early floods can destroy whole crops. This type of garden, used for both cash and subsistence crops, is mostly cultivated by *ribereño* peoples who live along the major white-water rivers (Amazon, Putumayo) and also provides an alternative for traditional groups living along the Caquetá River. The long-term instability of this form of gardening caused by unpredictable river levels thus affects the subsistence and market economies of less traditional groups as well as the alternative forms of horticulture of traditional groups.

The slash-and-burning of horticultural plots in secondary forest—which are less diverse than those in mature forest and less productive than those on alluvial soils—thus remains the most dependable gardening option. This type of gardening has become habitual for groups with restricted territories, and it is an alternative for both *ribereño* peoples and traditional groups of *terra firme*. The repeated slashing of secondary forest leads to soil degradation and increasing weed invasion.

Effects on Health

Most interviewees stressed outright the increased incidence of respiratory and intestinal (most likely viral) diseases in recent years, observing that the diseases were not only stronger than before but also of formerly unknown kinds. Vector-borne diseases like malaria or dengue fever apparently have not increased, but changing patterns of river fluctuation may improve the breeding conditions for vectors in still unpredictable ways.

Indigenous peoples are familiar with the seasonal appearance of viral diseases, which is closely related to river fluctuations, precipitation, and temperature, and they use both material and spiritual practices to prevent them. Illness and diseases, in indigenous peoples' view, live in the trees and in water. The rising waters and the winds of *friaje* are seen to dislodge diseases and flush them away. Failure of flooding and *friaje* winds is believed to cause contamination to accumulate in the atmosphere and in the waters, while rising temperatures are seen to worsen the accumulated contamination, causing respiratory and intestinal diseases to linger and develop into new, more-virulent kinds. Interviewees explicitly stated to us that ritual practices and preventive measures meant to counter the adverse effects and dangers of each season have increasingly failed to have the desired effect.

Heat, too, affects human health. High temperatures in open areas, such as garden plots, cause fatigue and headaches, increasing what Indians understand as the circulation of illness in nature.

Other effects on human health—not noted by interviewees, but predictable—have to do with the potential diminution of food supplies. Fish is the main source of protein and is one of the resources more directly hit by altered seasonality. Though manioc, the main source of carbohydrates, is a very resistant and adaptable crop, other crops that enrich and complement the diet are affected by the changing climate, as are alternative sources of protein, minerals, and vitamins such as wild and cultivated fruits and animals.

Collateral Benefits
The unpredictability of seasons has lessened the incidence of pest infestations. The months of September and October (northern regime) are known as the "time of worms," a season of fluctuating water levels and thunderstorms, when the forest fills with worms that feed on fruit. In recent years, the worm infestation has failed to materialize.

The great summer period is characterized in some regions (Caquetá and Putumayo basins) by the appearance of swarms of blood-sucking gnats. These are not known disease vectors, but they annoy people and cause infections on small children's skin and allergic reactions in some individuals. The failure of clearly distinguishable summer periods to occur has diminished the incidence of this plague.[7]

Gender, Age, and Social Differential Effects
Most of the above-mentioned changes have effects across genders and ages, but some groups are especially vulnerable.

Young children are the group most vulnerable to viral diseases, which, as shown above, are one of the effects of climate change most strongly felt by indigenous peoples. Also, crop failures and the shifting availability of protein can be expected to directly and irreversibly affect the growth and healthy development of young children.

Elders and ritual masters grow and use a large number of plant species for healing and ritual practices, but their ability to carry out these practices is compromised when they cannot successfully grow the plants they need. Also, new and more-virulent diseases confound their healing and prevention abilities.

Women carry a heavy burden of impact, as they traditionally bear the responsibility for routine labor in horticulture. Work in open fields under rising temperatures, crop failure, and the need to replant failed crops, together with lower yields, affect both women's physical health and their psychological well-being. Gardens in secondary forest or half-burnt gardens in mature forest take more work to eradicate weeds and manually complete the burning process. Children's ill health and malnutrition impose an additional burden on women as caregivers.

Young males, who bear the main responsibility for hunting and fishing (and hence for the supply of protein) and the clearing of forest for new garden plots, are affected by the falling availability of fish. The effort needed for hunting and fishing increases, and it becomes more difficult to judge the right timing for the slash-and-burning of new plots.

The indigenous peoples who have greater territorial autonomy—who derive their livelihood mostly from forest and water resources and maintain an active and engaged ritual life—see their livelihood most strongly affected. They place great value on gardens slashed in mature forest, planted with a high variety of species; depend heavily on fish and game for protein; and take care of their health with their own means and knowledge. Their livelihood rests on their ability to interpret regular natural cycles and act in accordance with them. Though they certainly have contact with mainstream society, are incorporated to some degree in the market economy, and have access to public health and education services, a large proportion of their livelihood depends on their knowledge, use, and management of forest and water resources. Our interviews showed these to be the indigenous peoples most aware of and most vulnerable to the effects of climate change and variability. In this category are the indigenous groups of the Vaupes area (Eastern Tukano, Maku-Puinave, and Arawak linguistic stocks), the Caqueta-Putumayo region (Witodo, Bora-Miraña, and Andoke stocks), the Guainia region (Maku-Puinave and Arawak stocks) and some groups of the Amazon Trapeze (Tikuna, Peba-Yagua, and Tupi stocks) (figure 2.7).

Figure 2.7   Colombian Amazon Population Density and Indigenous Peoples' Areas

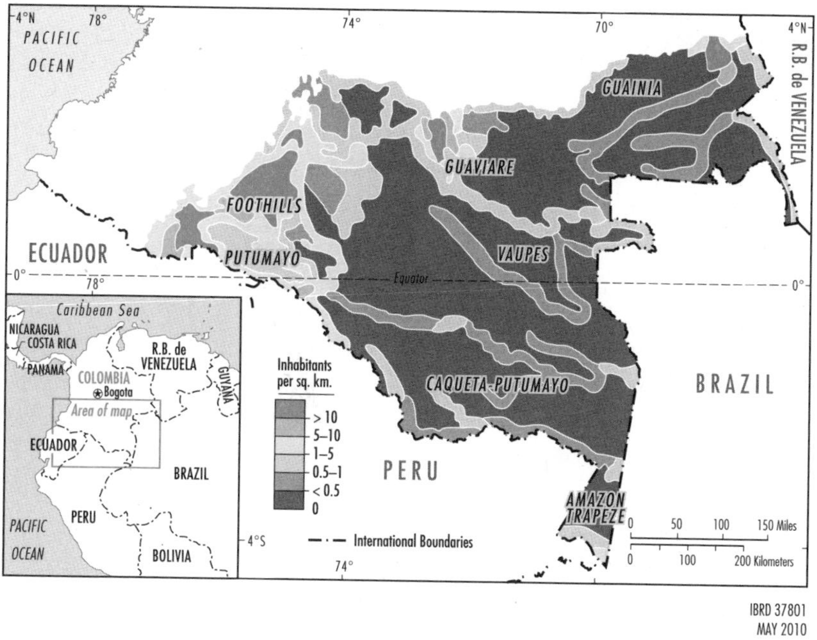

IBRD 37801
MAY 2010

Source: Adapted from El Espectador 1986.

A different scenario plays out for indigenous peoples who have restricted territorial autonomy and little or no access to mature forest and who depend on horticulture in secondary forest, cash crops from alluvial soils, commercial fishing, wage labor, tourism, and sales of handicrafts for their livelihood. They feel the effects of climate changes to the extent they use river and forest resources, and they are equally affected by diseases and rising temperatures. These indigenous peoples are less in tune with the seasonal calendar (or their awareness is restricted to those aspects that directly affect their activities), their traditional knowledge is more limited, and generally, ritual specialists play a lesser role in their lives. Their greater access to the market and to public health and education services provides them with resources to buffer the impacts on their livelihood that result from changes in the reliability of natural resources. They are less aware of climate change and variability and less vulnerable than more traditional groups, but the combined effects of climate change and variability tend to prompt or accelerate their reliance on wage jobs, market integration, and migration to urban areas, causing increased poverty. In this category are indigenous peoples on the fringes of the colonization areas at the foothills of the Andes and along the Guaviare River, and indigenous peoples of other regions living near urban areas (figure 2.7). Trujillo (2008) presents quantitative data on market dependence for their livelihood of indigenous groups living in the vicinity of the city of Leticia, along the Amazon River.

Transforming Structures and Processes

Particularly in the western part of the Colombian Amazon, indigenous peoples' livelihoods, and their options in responding to climate change and variability, depend closely on powerful social trends affecting their lives.

The eastern part of the Colombian Amazon region, with its few urban enclaves,[8] is mainly inhabited by indigenous peoples, and most of the central and eastern Colombian Amazon forests are fairly well preserved. But in the western and northwestern part of the region, deforestation is extensive and advancing rapidly (figure 2.8). The colonization of the western part began early in the 20th century and has continued since, pushing toward the eastern part of Colombia. This movement has been spurred by political violence, prompting people to leave violent areas, and by the appetite for land on which to grow illegal crops. This part of the region has cities connected by road to the main Colombian urban centers, a mostly non-Indian population, and cattle-ranching colonization.

Since the 1980s, a national policy of environmental and indigenous protection has brought about the creation of new national parks and Indian preserves or *resguardos*.[9] Indian preserves now total 247,000 km^2, and national parks and reserves cover 58,840 km^2.[10] Taking into account the several overlaps between Indian *resguardos* and protected natural areas, more than 70 percent of Colombian Amazonia is in principle under some protection (figure 2.9). Even so, the area is not safe from advancing colonization and destruction: the presence of the Colombian state is weak, Indian political institutions are incipient, and the region is affected by powerful actors and processes that threaten indigenous livelihoods just as much as climate change and variability do.

Colonization pressure from the west is strong and hardly contained by the presence of the extensive protected natural and indigenous areas. Armed conflict in Colombia displaces large numbers of people to the periphery and into the illegal economy. Western Colombian Amazonia has one of the largest areas of the world devoted to coca plantations. These plantations not only cause massive deforestation but also pollute the waters and affect indigenous populations who get involved in illegal

Figure 2.8   Colombian Amazon Forest Cover

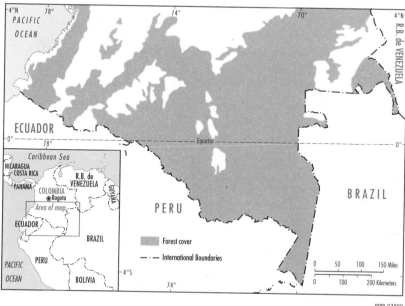

IBRD 37802
MAY 2010

Source: Adapted from IGAC 2002.

activities. The western Colombian Amazon is an historical stronghold and haven of the Colombian guerrillas, kept at bay from the main rivers and population centers only by the increasing presence of the Colombian military. The effects of illegal coca cultivation, guerrillas, and governmental military operations are felt more strongly in the west, where Indians have frequently been killed or displaced from their territories. Indigenous peoples in the larger eastern regions are now safer as the result of an increased military presence, but all these social and political processes are as unpredictable as temperature, precipitation, and seasonal shifts.

In the east, the more traditional indigenous peoples have territorial autonomy in the large expanses of Indian preserves, but the state offers them only limited protection from the sinister forces discussed above. Indians have, by law, the status of land authorities and can rule their territories according to their own cultural modes, but they lack the capacity to enforce or deter the unwanted presence of other actors or to contain social disintegration, which happens, for example, through their youths joining guerrilla forces or going to work in coca plantations. Currently, the

Figure 2.9   Indigenous Territories and Conservation Areas in Colombian Amazonia

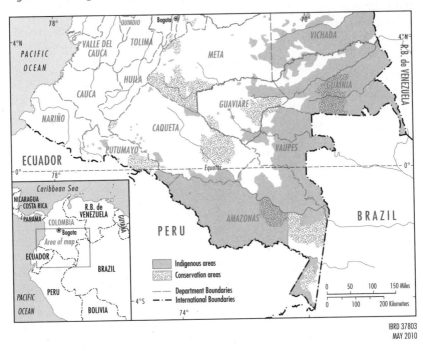

IBRD 37803
MAY 2010

Source: Adapted from Consolidación Amazónica-Colombia 2005.

indigenous federations play a limited role in protecting their lands. In Colombia as in Brazil, they have established alliances with environmental groups, which may be enhanced in the future and thus enable them to protect their lands more effectively.

New social factors affect the situation, too. Indians depend not just on natural resources for their livelihood; they also need education and health services and money to buy commodities (clothing, tools, and so forth) from the market. Money is scarce, and the need for it varies among groups—more intense among those who live on the fringes of colonization or urban areas, and less intense among groups with greater territorial autonomy. Besides occasional wage jobs or cash aid from the government or NGOs, the main source of money is the exploitation of natural resources: commercial fishing, timber extraction, production of handicrafts, gold mining, sale of wild fruits and game, and cash crops. These activities affect resource management and are seen to be among the causes of natural disorder: resources are extracted in excess; gold mining brings contaminated materials from the subsoil to the surface; and gardens are planted with a few crops intended for sale, rather than the diverse range of crops needed for ritual exchange and reproduction.

Indigenous people are increasingly integrated in the monetized economy. "We are blinded by money; the products of the forest are now turned into business," said a Witoto man from the Amazon River region. Money is needed for schooling—a key concern for indigenous peoples. Young people, often men, migrate out of the region to work in coca fields or in timber camps, to make a sufficient income to raise a family. Climate change and variability affects the yield of gardens, and the availability of protein declines, prompting people to migrate. "Before, it was individuals who migrated; now it is entire families who move out of the *resguardo* areas," a Nonuya man from the Caquetá River declared.

Institutional Framework

The institutions most trusted by indigenous peoples are their own cultural institutions—not the formal or political indigenous organizations, but the traditional modes of organization and coordination among people and with the natural elements. Ceremonies and rituals vary among groups, but all share the purpose of healing and maintaining nature and society. Examples of such rituals include the ceremony of *ayahuasca* (*Banisteriopsis caapi*, an entheogenic vine) among peoples of the Andean foothills in the western part (Schultes and Raffauff 1994); ritual exchanges and dance festivals among *malocas* (longhouses)[11] and the ritual consumption of tobacco and coca in the Vaupes and Caquetá-Putumayo regions (Echeverri 1997; Kronik 2001); and the exchange of manioc beer to carry out communal works among the peoples of the Amazon Trapeze. All these practices depend firmly on territorial autonomy and on the ability to produce and exchange abundant food. The situations of the leaders and maintainers of these institutions are very diverse. Some indigenous political organizations work well for the indigenous peoples, but some do not coincide with the formally recognized indigenous "authorities" and political organizations. In many cases, financial support and aid that is meant for indigenous groups reaches only the formally recognized authorities.

Indigenous peoples in *resguardos* have formal authorities and organizations, which have mainly representative functions before the state. They also have their natural, traditional authorities, whose political capacity is rooted in ritual activities and closely linked to their ability to maintain and reproduce life. Their main engagement is through constant dialogue with nature and through dialogue and exchange among people.

Indigenous political organizations are often allied to and in mutual dependency with NGOs. The first obligation of indigenous political organizations is to be the direct representatives of communities at the grassroots level, which is not an easy task, particularly considering the partial militarization of the region.

NGOs provide valuable outside support. Their work is appreciated because they are willing to reach out to indigenous peoples and flexible in taking action and making decisions. However, in many cases their action is ephemeral because their resources tend to be tied to projects. Moreover, they may generate conflicts and jealousy with government officials for providing services that should have been extended by state institutions. A few NGOs (financed by international agencies) have maintained close and continued support for indigenous initiatives, which has helped to reinforce indigenous traditional institutions.

National and even regional institutions generally have a limited presence in indigenous territories. The Directorate of Ethnic Affairs of the Ministry of the Interior is mandated by law to conduct consultation meetings with indigenous peoples for the execution of projects in their territories. The *Direccion de Etnocultura* of the Ministry of Culture conducts a program that gives financial support for the construction of *malocas*. But insufficient understanding of indigenous peoples and their cultures and the centralized management from Bogotá may, in fact, produce undesired effects when funds are injected without including the social and religious meanings of the *maloca* institution. Another program, the *Guardabosques* (forest keepers) of the Presidency, is a strategy in the war against drugs; it offers cash benefits to families that commit not to grow coca and to manually eradicate existing plantations. Yet, the program could potentially compromise the cultural practices of peoples who grow and use coca as part of their traditional life.[12] In the Amazon Trapeze, however, we learned of a displaced Witoto community that has managed to negotiate with the program to maintain its traditional coca fields and to employ program resources to fund innovative solutions.[13] The Unit of National Parks of the Ministry of the Environment, which is in charge of the custody and management of national natural parks, has historically been in conflict with indigenous peoples, particularly in parks that overlap with indigenous territories. However, in recent years, the unit has made efforts to improve relations with indigenous peoples and work on concerted management plans. Finally, Instituto de Investigaciones de la Amazonia (SINCHI), the official Amazonian research institute with headquarters in Leticia, carries out programs on alternative crops and promotes the use of new cultivation methods, although with limited success thus far.

Most regional institutions—from the *departamentos* (administrative territorial units) and municipal authorities, health and education services, to the official corporations in charge of the control and management of the natural resources of the region—often have insufficient knowledge of Indian life, culture, and needs, impairing their relationship with indigenous peoples.

Adaptation and Survival Strategies

Indigenous peoples are accustomed to climatic changes and variations. One of their main adaptation strategies is multiactivity—applying a varied array of skills to do many different things in different ecological environments. People are fishers, planters, hunters, know the river and forest, and also learn new techniques from nonindigenous people and use them to their benefit. This multiactive capacity is a great cultural asset, common to all indigenous groups, that allows them to adapt to and cope with changes. As one Cocama man from the Amazon River put it, "We need to learn to do everything in order to survive." Nevertheless, stressful climatic extremes put these abilities to the test, and indigenous peoples are now searching for solutions to imminent threats—both climatic and social—to their livelihood.

Horticulture

One of the most fundamental livelihood systems now at risk is horticulture, particularly in gardens opened in mature forest. "We no longer expect the dry season," stated a man from the Caquetá River. The lack of a clearly distinguishable dry season has led people to explore several alternatives: (1) to slash the forest and burn as soon as a few dry days take place; (2) if this does not happen, to manually build fires and proceed to burn the garden area piecemeal; (3) to convert to open gardens in secondary forest, which requires shorter dry periods; and (4) to open slash-and-mulch (with no burning) gardens on alluvial soils, with the risk of losing the crops if the river rises before plants are ready to be harvested. The latter systems support fewer species, as noted above, and they are best suited for the sweet varieties of manioc, which grow faster than the bitter ones. For the more traditional indigenous groups of the northeastern Colombian Amazon, these adaptations result in loss of crop diversity and affect ceremonial life. Among *ribereño* indigenous groups and those with less territorial access, these changes have been going on for some years. It would appear that traditional slash-and-burn Amazonian horticultural systems in mature forest, as reported by ethnographers, will soon disappear and that many groups will adopt less ecologically dependent, and less species-diverse, systems.

Another change in horticultural practices is that of prolonging the life of garden fields. Traditional indigenous peoples' gardens do not last more than a few years and are then abandoned to fallow. Amazonian peoples do not customarily practice soil enrichment. SINCHI has been promoting new techniques in the Amazon Trapeze and in the Igaraparana River (Caquetá-Putumayo region): instead of burning, it advocates maintaining the productivity of existing fields by adding natural fertilizers (ashes, rotted wood, and so forth). This system is just being introduced, and its results are still to be seen. It requires a much more intensive dedication to horticulture, which reduces the time that people have for other subsistence activities.

Indigenous Communities' Perceived Needs for Help

Though most of the destructive trends originate outside the indigenous communities, what most concerns indigenous peoples is how they themselves become involved and transformed by them. A recent meeting among elders of the Muinane and Nonuya tribes in the Caquetá River area addressed two conjoining trends impinging on their lives: problems with production (horticulture, fishing, game) and social problems (guerrillas, illicit trade, resource exploitation, outmigration, increasing health problems). Their assessment of causes and their proposals for action clearly illustrate indigenous peoples' current quandaries.

The elders noted that they are not practicing the ritual control of nature and society that was ordained by the Creator to mitigate and reverse the effects they now witness. Natural disarray is a reflection of social disarray. Reversing this development is not the task of a single person; it requires coordination among ritual specialists of different tribes. The formal indigenous authorities that have been put in place (elected communal chiefs and indigenous political leaders) do not have adequate knowledge and capacity to deal with this situation, because they are following what the elders refer to as the "white people's" agenda and mode of thinking. It therefore falls upon elders and household heads to resume their moral responsibilities. The agenda and concerns of mainstream society are also increasingly dominating indigenous peoples' time and space that would otherwise be devoted to ritual dialogue and practice and to the traditional education of younger generations. Efforts need to be made to control and curtail these harmful interferences. The greatest danger comes from what some indigenous peoples of the Amazon refer to as the "Hot and Poisonous World" that comes from the subsoil, where it should be kept. Released by extracting oil over the past decades, it now reaches indigenous peoples in the form of money, weapons, luxuries, alcohol, media, and diseases—and has also resulted in contamination of the air that surrounds them and ought to support seasonal rhythms. "If all these changes are the result of a planetary disorder, what can we, a little group of people, accomplish?" the group of elders wondered.

The elders posed a shared-responsibility proposal. They need outside support and mediation to publicize their own diagnosis and proposals, based on lessons from their own systems of natural management. But this relationship with outside institutions needs to use a new approach: not accepting and receiving proposals conceived by outsiders, as has been the rule, but searching for help and support for their own agenda and addressing their own concerns.

Summary

Climate change and variability increase pressure on natural resources as people search for sources of income. It also undermines the ability of traditional leaders to exercise effective social control because it challenges their ability to predict and control natural cycles and, hence, human well-being.

For indigenous populations of the Colombian Amazon, changes in the timing of the dry and rainy seasons as well as alterations in the flood pulses of rivers, winds, and heat have become apparent since 2000 and increasingly since 2005. Indigenous peoples are keen observers of natural rhythms and have accumulated a large and sophisticated knowledge of annual seasonal cycles. Their traditional livelihood systems (slash-and-burn horticulture, fishing, hunting, and gathering of wild resources) are closely tied to predictable and well-established seasons. Indians know and are able to interpret complex ecological indexes of the timing of seasons, which were—up until a decade ago—clearly established and well known to them. These natural rhythms ordain the interrelation of water, wind, heat, fish, animals, insects, wild fruits, and human productive and ritual activities and serve to regulate, defend, and maintain life. According to key people interviewed in different areas, the natural signs and indexes they now perceive are "alarming": for example, seasons are occurring outside their normal time, the regular flood and ebb of rivers is out of synchrony with the fall of wild fruits on which fish depend for food, and heat is increasing. These changes have direct impacts on livelihoods, mainly on horticultural activities, crops, and reproduction of fish, and on human health.

The greatest current concern of indigenous peoples in the Colombian Amazon region, however, is their social situation. The traditional harmony of their lives with nature is disturbed by climate change and variability, but also by the effects of advancing colonization, destruction of the forest, political unrest, illegal coca cultivation, excessive resource exploitation, gold mining, and trade. In combination, these changes lead to increasing social disarray.

The effects of global climate change and variability should be assessed together with the effects of these other trends and processes. Support to contain deforestation and to uphold traditional indigenous peoples' institutions should be mobilized. Traditional indigenous peoples' institutions depend highly upon the vitality of their knowledge systems and, therefore, upon the use and dissemination of their languages. In areas with strong urban and market influence, and elsewhere where indigenous languages are at particular risk, programs that retain these languages have great value, as much knowledge is contained in a people's language.

Notes

1. An overview of indigenous diversity in the Colombian Amazon can be found in Instituto Colombiano de Antropología (1987) and its companion volume Correa (1993), which discusses contemporary cultural and social changes. More recent assessments of the social and territorial situation of these indigenous groups are found in Franky (2004) for groups of the Vaupes area, Fundación Gaia-Amazonas (2004), García and Ruiz (2007), Roldan (2002), and Vieco, Franky, and Echeverri (2000).

2. Here and in other places, indigenous peoples' perceptions of increased heat coincide with meteorological observations. However, this does not mean that perceptions present evidence of climate change; nor are they intended to do so.

3. The difference in fluctuation reflects the difference not only in seasonal regimes but also in the overall atmospheric-oceanographic processes. Further south, there is a strong influence of El Niño and the Pacific Ocean; further north, the influence is weaker, and other processes (linked in part to the Atlantic) dominate.

4. Marengo and others (2008) assert that the 2005 drought was not linked to El Niño, but to warming sea–surface temperatures in the tropical North Atlantic Ocean.

5. Paoli, Curran, and Zak (2006), among other studies, document the way that the variation in the fruiting of tropical trees can anticipate weather changes.

6. Other factors also influence agriculture—for example, shortening of fallow times because of greater population pressure or greater market demand. People might genuinely believe that it is only the climate that is changing, or they might note the increase in the percentage of open areas (and the decrease of forest) and the associated greater perceived warmth, or some local decline in precipitation due to reduction in vegetation cover. These possibilities are not mutually exclusive; they simply emphasize that climate change and other factors can coincide.

7. Andersen, Geary, Pörtner, and Verner in Verner (2010) addresses health implications of climate change in LAC.

8. San Jose del Guaviare (pop. 50,000), Leticia (30,000), Mitu (7,000).

9. *Resguardos:* lands titled to indigenous populations.

10. Indian preserves make up 62 percent of the Colombian Amazon, and national parks and reserves make up 15 percent.

11. A *maloca* is an ancestral common house—by some referred to as a longhouse. The communal center of the *maloca* can be 10 × 10 × 10 meters, although each community has a *maloca* with its own unique characteristics. Several families with patrilineal relations live together in a *maloca*, distributed around the longhouse in different compartments. During festivals and in formal ceremonies, which involve dances for males, the longhouse space is rearranged; the center of the longhouse is the most important area where the dance takes place (Kronik 2001).

12. Caquetá-Putumayo and Vaupes regions.

13. This community was originally from the Caquetá-Putumayo region.

References

Andersen, Lykke, John Geary, Claus Pörtner, and Dorte Verner. 2010. In *Reducing Poverty, Protecting Livelihoods and Building Assets in a Changing Climate: Social Implications of Climate Change in Latin America and the Caribbean,* ed. Dorte Verner. Washington, DC: World Bank.

Bunyard, Peter. 2008. "Why Climate is Dependent on Biodiversity." In *Fronteras en la globalización: Localidad, biodiversidad y comercio en la Amazonia,* eds. C. Zarate and C. Ahumada, 21–42. Bogotá: Fundación Konrad Adenauer, Pontificia Universidad Javeriana.

Consolidación Amazónica-Colombia. 2005. "Areas protegidas y territorios indígenas en la Amazonia colombiana." http://www.coama.org.co/otros/mapas/TIs_APs_Col_Carta.jpg.

Correa, François, ed. 1993. *Encrucijadas de Colombia Amerindia.* Bogotá: Instituto Colombiano de Antropología, Colcultura.

Echeverri, J. A. 1997. "The People of the Center of the World—A Study in Culture, History, and Morality in the Colombian Amazon." Unpublished dissertation, Graduate Faculty of Political and Social Science, New School for Social Research, New York.

El Espectador. 1986. "Así es Colombia." Editorial, *Periódico El Espectador* (Bogotá).

Franky, Carlos. 2004. "Territorio y territorialidad indígena. Un estudio de caso entre los tanimuca y el Bajo Apaporis (Amazonia colombiana)." Tesis de Maestría. Universidad Nacional de Colombia, sede Leticia.

Fundación Gaia-Amazonas. 2004. "Pueblos indígenas del noroeste amazónico: Realidades y mundos posibles. Seminario internacional. Parte 1." Bogotá: Ediciones Antropos. http://www.coama.org.co/documentos/articulos/PI-NWA_Parte1.pdf.

García, Paola, and Sandra Lucía Ruíz. 2007. "Diversidad cultural del sur de la Amazonia colombiana." In *Corpoamazonia, Diversidad biológica y cultural del sur de la Amazonia colombiana: Diagnóstico*: 259–306. Bogotá: Ramos López Editorial Fotomecánica Ltda. http://www.corpoamazonia.gov.co/Planes/download/biodiversidad/AMAZONIA_C3.pdf.

IDEAM (Instituto de Hidrologia, Meteorologia y Estudios Ambientales). 2008. The Colombian Institute for Hydrology, Meteorology, and Environmental Studies.

IGAC (Instituto Geográfico Agustín Codazzi, Colombia). 2002. "Cobertura vegetal y uso." http://ssiglims.igac.gov.co/ssigl/mapas_de_colombia/galeria/IGAC/CoberturaVeg.pdf.

————. 2008. "Mapa físico-político de Colombia." http://ssiglims.igac.gov.co/ssigl/mapas_de_colombia/galeria/IGAC/Matis_Colombia.pdf.

Instituto Colombiano de Antropología. 1987. *Introducción a la Colombia Amerindia.* Bogotá: Editorial Presencia. http://www.lablaa.org/blaavirtual/antropologia/amerindi/index.htm.

INEC (Instituto Nacional de Estadística). 2001. *VI Population Census.* Bogotá, Colombia.

IPCC (Intergovernmental Panel on Climate Change). 2001. *Special Report on Emission Scenarios.* Geneva: IPCC.

————. 2007. *Climate Change 2007: The Physical Science Basis.* Contribution of Working Group I to the *Fourth Assessment Report* of the IPCC. Geneva: IPCC. http://www.ipcc.ch.

IWGIA (International Work Group for Indigenous Affairs). 2008. *The Indigenous World Yearbook.* Copenhagen: IWGIA.

Kronik, Jakob. 2001. "Living Knowledge—Institutionalizing Learning Practices about Biodiversity among the Muinane and the Uitoto in the Colombian Amazon." PhD dissertation, Roskilde University Center, Roskilde, Denmark.

Layton, Heather Marie, and H. A. Patrinos. 2006. "Estimating the Number of Indigenous People in Latin America." In *Indigenous Peoples, Poverty, and Human Development in Latin America*, eds. Gillette Hall and H. A. Patrinos. New York: Palgrave.

Marengo, J. A., C. A. Nobre, J. Tomasella, M. F. Cardoso, and M. D. Oyama. 2008. "Hydro-Climatic and Ecological Behavior of the Drought of Amazonia in 2005." *Philosophical Transactions of the Royal Society* B (363): 1773–1778. http://journals.royalsociety.org/content/238x818l0815588k/fulltext.pdf.

Orlove, B. S., J. C. H. Chiang, and M. A. Cane. 2002. "Ethnoclimatology in the Andes: A Cross-Disciplinary Study Uncovers a Scientific Basis for the Scheme Andean Potato Farmers Traditionally Use to Predict the Coming Rains."*American Scientist* 90: 428–435.

Paoli, Gary D., Lisa M. Curran, and Donald R. Zak. 2006. "Soil Nutrients and Beta Diversity in the Bornean Dipterocarpaceae: Evidence for Niche Partitioning by Tropical Rain Forest Trees." *Journal of Ecology* 94 (1): 157–170.

Roldán, Roque. 2002. "Territorios colectivos de indígenas y afroamericanos en América del sur y central: Su Incidencia en el desarrollo." Paper presented at conference of Banco Interamericano de Desarrollo Departamento de Desarrollo Sostenible, "Desarrollo de las Economías Rurales en América Latina y el Caribe: Manejo Sostenible de los Recursos Naturales, Acceso a Tierras y Finanzas Rurales," Fortaleza, Brazil, March 7, 2002. http://www.pueblosaltomayo.com/articulos/tierras-y-territorios/territorios%20colectivos_BID.pdf.

Salick, Jan, and Anja Byg. 2007. *Indigenous Peoples and Climate Change*. Oxford, U.K.: Tyndall Centre for Climate Change Research.

Schultes, R. E., and R. F. Raffauff. 1994. *El bejuco del Alma—medicos tradicionales de la Amazonia Colombiana*. Bogotá: Uniandes.

Trujillo, Catalina. 2008. "Selva y mercado: Exploración cuantitativa de los ingresos en hogares indígenas." Tesis de Maestría en Estudios Amazónicos. Universidad Nacional de Colombia, Sede Amazonia, Leticia.

Verner, Dorte. 2010. *Reducing Poverty, Protecting Livelihoods and Building Assets in a Changing Climate: Social Implications of Climate Change in Latin America and the Caribbean*. Washington, DC: World Bank.

Vieco, J., C. Franky, and J. Á. Echeverri, eds. 2000. *Territorialidad indígena y ordenamiento en la Amazonia*. Leticia: Universidad Nacional de Colombia Sede Leticia, Instituto Amazónico de Investigaciones Imani, Programa COAMA.

CHAPTER 3

Indigenous Peoples of the Andes

The impacts that climate change and variability may have on the montane ecosystems of LAC and the indigenous peoples who inhabit them have been little studied. Alterations in the climate may affect people's means of subsistence, such as arable farming and livestock rearing, both positively and negatively. New threats and possibilities may force or urge people to change their customs and traditional forms of production as well as their community relationships. In the longer run, these changed forms of production may jeopardize the people's food security and affect their health. Indigenous peoples are highly vulnerable to such changes, given their high dependence on their traditional knowledge for managing the environment. Moreover, their highly location-based identities make the prospect of migration particularly threatening to them. Their oral transmission of knowledge creates a deep historical cultural memory, but their location-based beliefs and rituals create a strong attachment to landscape features for cultural and religious reasons. Unpredictable changes in climatic conditions may render indigenous knowledge useless or unsuitable, which may in turn threaten not only sources of sustenance but also the development and maintenance of their culture. These concerns, as well as whether and how indigenous peoples in the Andes regions are adapting to these changes, are the subject of this chapter. Naturally, adaptation is related to types of impacts and vulnerability, which vary depending on cultural features, social capital, productive practices, and socioeconomic and political situations—all of which influence people's resilience and adaptive capacity (Verner 2010).

Of the Andean countries, Bolivia has the highest proportion of indigenous peoples (table 3.1). Focusing on Bolivia, with secondary information from other Andean countries, particularly Peru, this chapter analyzes the social impact of climate change and variability on indigenous peoples' livelihoods and social well-being. In Bolivia, all of the Andean ecozones and systems are present. In addition to its rich cultural variation, Bolivia is also known for its recent indigenist- and workers-unions-led government. Choosing Bolivia as the main entry point is meant to contribute insights comparable with current processes and conditions in other Andean countries. The first section outlines the threats from climate change in the Andean regions, noting how these threats have begun to affect indigenous people. The second and third sections present case studies from two areas of Bolivia—the northern Altiplano and the northern Yungas.[1] The fourth section concludes.

Threats from Climate Change

In the Andean regions, the effects of climate change and variability are mainly manifested through the rapid retreat of glaciers; increasing frequency of the phenomenon of El Niño and La Niña, often causing widespread damage;[2] increases in mean temperatures; an increase in the thermal daily amplitude; and more intense and frequent anomalies in sea-surface temperature in the Pacific Ocean.[3]

Table 3.1   Facts on the Indigenous Peoples of the Andean and Sub-Andean Countries

Country	Number of peoples	Approximate number of indigenous individuals (1000)	Percent of total population
Bolivia[a]	36	5,200	62.0[*]
Chile[a]	8	1,060	6.5
Colombia[a]	92[d]	1,400	3.4
Ecuador[a]	14	830[e]	6.8
Peru[b]	65	8,800	33.0
Venezuela[c]	40	570	2.2

Sources: Authors' elaboration based on following sources: (a) national census self-identification; (b) Layton and Patrinos 2006; (c) by self-identification (IWGIA 2008); (d) Instituto Socioambiental (ISA); and (e) INEC 2001.
Note: (*) Older than 15 years of age.

Warming in high mountain regions of the South American continent melts glaciers, snow, and ice, affecting farming and the availability of water to coastal cities, power generation, and tourist activities. Outbursts of glacial lakes also pose a threat to lives and livelihoods. Except for brief periods, all tropical glaciers lost considerable mass over the past century (Georges 2004; Gil 2008; Kaser 1999), and the surface area covered by glaciers in Bolivia and Peru shrank by nearly 30 percent between 1970 and 2006. Looking ahead, the average global warming for 1990–2100 is projected to be between 1.4 and 5.8°C, but warming is projected to be much greater at higher elevations (IPCC 2007a). At present rates, several of the Andean glaciers may vanish in a few decades or sooner. IPCC (2007b) warns that "changes in precipitation patterns and the disappearance of glaciers are projected to significantly affect water availability for human consumption, agriculture, and energy generation."

Rising mean temperatures also affect agriculture in other ways. Plants (including crops and pasture cover) require more water at higher temperatures because of higher rates of evapotranspiration. Lower elevations may become too hot and dry for customary crops. As mountain pastures dry up, or are substituted by competing land uses, they cannot support livestock, and there is an increased risk of erosion of the soils of headwater basins. Meanwhile, some higher-altitude areas that used to be too cold for agricultural production have now become warm enough to be productive, improving some people's livelihoods. Widely reported for Peru is the upward movement of crops: maize and, in a few areas, potatoes can now be grown at higher elevations.

Indigenous populations in the Andean regions tend to be socially, politically, and economically marginalized, and their livelihoods depend on natural resources, making them highly vulnerable to climate change and variability (box 3.1). Field research shows they are being affected by a long list of negative impacts: growing scarcity of water; erosion of ecosystem and natural resources, for example, through salinization of soils; changes in biodiversity as a consequence of the spread of alien species; plant diseases affecting crops; a higher death toll among livestock; higher risks of infectious diseases; and crop losses.

These effects lead to material and human losses—deaths and loss of property—and threaten food security, both within indigenous villages and among the populations who depend on the food produced by these villages. Likewise, these losses induce migration toward the cities, where people crowd into poor, conflict-ridden neighborhoods on urban fringes, creating the need for government solutions. A much-larger-scale parallel

Box 3.1

Glacial Retreat Gravely Affects Highland Herders

Dozens of villages of Quechua-speaking alpaca herders are located close to glaciers in the southern highlands of Peru. Wool and meat from alpacas have been
the basis of the herders' economy for centuries, whether bartered for other food-
stuffs or sold for cash. Pasture is abundant in this region during the rainy season,
but dry-season pasture is limited to the areas that can be irrigated from streams
that receive glacial meltwater.

In recent decades, the glaciers have been shrinking visibly, the streams have less
flow than before, and many pastures have dried up altogether. The herders are
deeply concerned at the rapid pace and the apparent irreversibility of these changes.

They fear that they will have no choice but to leave their villages permanently
and that their communities will disappear altogether. Though they frequently travel to barter and sell their wool and meat, and sometimes work for months or years
outside the villages to accumulate capital to purchase animals, they always have
retained ties to the villages where their kin live. The abandonment of these villages
is deeply distressing to them. A minority of them hope that simple projects—
improved veterinary care for their animals, support of irrigation projects—will
delay the decline of the herds, but glaciologists' projections confirm the pes-
simism of the majority: dry-season stream flow will decline to a tenth, or less, of
recent levels. Moreover, the herders view the mountains as the home of spirits
with whom they interact regularly in prayer and through sacrifices (which consist
chiefly of burning small quantities of animal fat and plant matter). They comment
that the mountains are saddened by the rapid retreat of the glaciers and recall old
myths that the world will end when the tallest mountains lose their ice.

Source: Orlove 2009.

process of partly planned extensive migration to the lowlands also has
consequences for vulnerability to climate change and variability. There
may be more water in the lowlands, but people are more dependent on
the market economy there.

Based on secondary and field information about perceptions, in the
two case studies that follow we analyze the effects of climate change
and variability on indigenous populations in two parts of Bolivia. Of
Bolivia's population of nearly 10 million, 62 percent is indigenous
(table 3.1), and more than half of them live in rural areas. Bolivia's

great diversity of cultures is distributed among 39 different indigenous peoples, of whom 31 live in Amazonia, 3 in the Chaco, and 5 in the Andes (figure 3.1). The Andean indigenous populations are much larger than those of the lowlands.

The case studies focus on indigenous people living in different regions of the Bolivian department of La Paz. The first of these is the Andean region of the Altiplano between 3,810 and 4,200 meters above sea level in the Titicaca Lake basin; it is home to the Aymara indigenous people. The second region is the sub-Andean region descending from the summit at 4,660 meters to the Yungas at 1,240 and 1,760 meters; this is home to the Aymara descendants of migrant Altiplano Aymara and Afro-Bolivian communities.[4] The indigenous people who live in the two regions studied rely

Figure 3.1   Major Indigenous Peoples in Bolivia

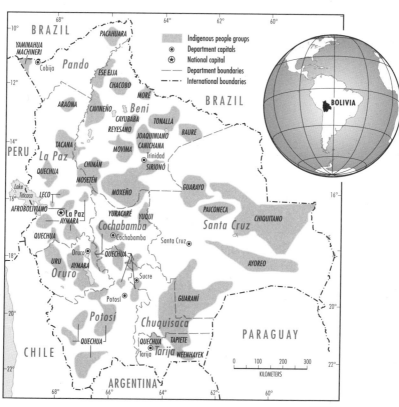

IBRD 37804
MAY 2010

Source: CIDOB 2007.

on arable farming and livestock production for their livelihood. Rainfall in the two regions is concentrated between December and March, and the dry period is from May to July.

Andean Region—Bolivia's Northern Altiplano

The Altiplano is a plain located between the Cordillera Oriental and Cordillera Occidental mountain ranges. It slopes gradually downward from north to south, varying in altitude from 4,115 to 3,665 meters. It presents a gradual decrease in moisture from north to south. The subhumid zone in the north has eight humid months a year, and the arid zone in the south has two to four humid months.

The case study area is the most densely inhabited zone in the Altiplano, where pre-Hispanic cultures such as the people of Tiwanaku and the Incas developed. It is located in the Puna Norteña ecoregion and the Puna Húmeda sub-ecoregion of the northern Altiplano, in the largest subbasin of the Titicaca Lake (figure 3.2). The landscape is mostly plains with some hills, and the vegetative cover consists of tough perennial bunch grasses, low and resinous bushes, and the remains of forest. Wetlands are found azonally. Land is mainly used for agriculture and livestock, but there is also mineral exploitation and some tourism.

Figure 3.2   Andean Region of the Northern Altiplano (meters above sea level)

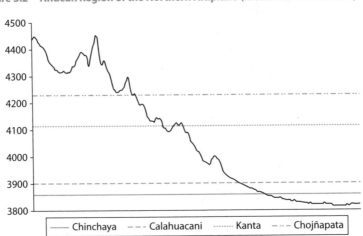

Indigenous Peoples

The Aymara, including the Quechua, is the main group among the five indigenous peoples who inhabit the Andean regions. They reside in the more densely inhabited area of the Altiplano, the mountain range, the valleys, and the Yungas, as well as the plains, as a result of colonization and migration. While there are Aymara in Peru and Chile, the majority of the Aymara live in the department of La Paz, principally the cities of La Paz and El Alto.

Livelihoods

The Aymara peoples' principal means of livelihood in the northern Altiplano is agriculture, including both arable crop and livestock production. Depending on the crop, sowing takes place between May and December. Farmers practice crop rotation, tilling, and soil fertilization with cattle dung. The better-off or better-organized communities use irrigation. Most of the communities use agrochemicals to fight pests, especially in onion and potato crops. The production is used for their own consumption and for sale at local markets. Livestock such as sheep and cattle are found at low latitudes, and llamas and alpacas are kept around 4,200 meters. Cheese from cow's milk, wool, meat, and livestock are sold at the local markets.

Governance—Transforming Structures and Processes

Transforming structures. Community members perceive the most important organizations to be the Campesino Workers' Union (Confederación Sindical Unica de Trabajadores Campesinos de Bolivia, CSUTCB)[5,6] and its central and subcentral offices,[7] which are the basis of the indigenous campesino workers' organization at the national level. The union represents the communal and intercommunal government; it allows for important decisions to be made in the communal assemblies regarding organization of productive and social life, handling of community issues, regulation of internal relationships, resolution of land issues, administration of justice according to the traditional Andean codes, and relationships with the regional authorities. The recent dramatic changes in the Bolivian government, with the coming to power of a combined labor union and indigenist movement (MAS) have further increased the importance of the agrarian union and its subsidiaries.

Conflicts between the agrarian union and the traditional *ayllus* seem rare.[8] The community authorities hold office for a year and rotate in a system that, though structured as elections, shows strong elements of traditional culture. In places where the traditional authorities survive, the rotation of office is often a mandatory statute of customary law, and it follows the agricultural cycle.

Processes. The Bolivian Agrarian Reform of 1952 freed the natives in the Altiplano from the obligation to provide free services on the land of the estate owners, and allocated up to 50 hectares of land to each native family. This caused a fragmentation of agricultural property, so that today, there are community members who own very narrow strips of land that allow production on only a very small scale.

To compensate for their shortage of land and diversify income-earning opportunities, people migrate either temporarily or permanently. Temporary migration, mainly to cities such as La Paz and El Alto, is driven by the need for economic survival. More permanent migration takes place mainly among young people, who leave to study or to work and hence improve their living standard and social status. Some of them return on a seasonal basis to help their parents during the harvest. There are also cases of parents with double residency, both in the community and in El Alto or La Paz.

Since 2007, some municipalities (for example, Ancoraimes) have prepared documents under Bolivia's National Climate Change Adaptation Program, which describes a "Strategy of Adaptation to Climate Change."[9] The documents include components of territorial planning, water safety, reinforcement of the productive system, organizational development, proactive initiatives in the health system, environmental protection, protection of health, and civil organization and participation. However, the communities we studied were not aware of the existence of this strategy.

Indigenous people in the area studied depend to varying degrees on social institutions and networks. They often maintain social and economic ties between different groups, and in many places they still support systems of food and labor sharing, including exchange, reciprocity, barter, or local markets. Such sharing systems have a role to play as adaptation strategies to environmental variability and stress. In the future, they could gain importance if adverse impacts of climate change and variability increase people's dependence on resources not available locally. The exchange systems may be threatened or disappear, however, as certain groups become more disadvantaged than others (Salick and Byg 2007). To supplement local or regional exchange practices, indigenous and traditional peoples might become more reliant on aid provided by the state, NGOs, or international organizations, especially in times of crisis.

Moreover, extensions of their social network across the country they live in, or even beyond, could become more common as an additional adaptation strategy to reduce socioeconomic vulnerability. For example, families that can count on members who migrate temporarily or permanently to work abroad may be more resilient to adverse climatic impacts than families whose members all remain within the community.

Effects of Climate Change in the Communities Studied

When asked about climate change, members of the Aymara communities in the highlands described many processes of change in the Altiplano that have occurred since the start of the century. In particular, they pointed to the increase in mean temperature, melting of the glaciers, and the absence of water (box 3.2). They referred to the change in seasons, increases in the intensity of rains and temperature, hailstorms and frosts during unusual times of the year, and droughts. Some also noted that the higher-elevation zones that were previously too cold for arable farming have become productive and now support the cultivation of hardy crop varieties.

Box 3.2

Quechua-Speaking Populations and Water Resources in the Cordillera

Climate change has begun to place stress on the Quechua-speaking communities of the middle slopes of the Cordillera Blanca of north-central Peru, even though these communities are among the highland Andean populations with the most secure and abundant access to water. The Cordillera Blanca contains not only Peru's highest peak, Huascarán at 6,768 meters, but also its largest contiguous territory above 5,000 meters, an area that receives immense deposits of snow each rainy season. A number of permanent streams flow down from this high country to the Río Santa, which descends from its upper course at 3,700 meters to 2,700 meters before entering a narrow gorge and flowing down through the coastal desert to the Pacific Ocean.

Quechua-speaking agriculturalists raise maize, wheat, potatoes, and other crops on the broad terraces. In addition, they have small herds of cattle and sheep. They also manage household and community woodlots of eucalyptus and pine. These woodlots benefit from climate change, as trees are now planted at higher elevations than they were in recent decades.

(continued)

Box 3.2 *(continued)*

Water is quite abundant in this region. Visitors from other parts of the Andes comment on the lushness of the vegetation that grows in the steep, uncultivated portions of the Río Santa canyon. The communities tap the permanent streams for irrigation. The irrigation water allows them to maintain rich pastures at higher elevations, so households can support small herds while keeping their arable lands—at a premium in this steep landscape—under crops. It also permits them to grow two crops a year at lower elevations. They can irrigate potatoes and maize in the dry season, months before the rains start, and thus they harvest their crops much earlier than other growers, while prices are high.

It is with water resources, though, that the negative impacts of climate change are most evident. Competition for water is growing; urban demand is one source. Populations are growing in Spanish-speaking mestizo towns along the main highway that follows the Río Santa. Recent decentralization programs in Peru give these towns greater control of the revenues that they receive. Some have used these funds to build water projects that tap the permanent streams high above the communities, carrying their water straight down to the towns past the communities that formerly used it. New economic activities also require water. Greenhouses are scattered across the valley floor, filled with water-thirsty bushes whose roses are cut, shipped in trucks, driven to the international airport outside Lima, and placed in the cargo vaults of trucks in a span of less than 12 hours. Hostels and lodges have been built in the valley to serve the growing population of tourists who hike and climb in the Cordillera Blanca; these tourists demand showers and flush toilets and prefer to stay in residences with well-watered gardens and swimming pools.

The rural communities themselves contribute to the increased water demand. Some have expanded their irrigation facilities, concerned that their lower lands need more water as temperatures rise or are simply aware that the towns—or other communities—might tap their streams. The long history of water abundance permitted the development of a system in which several communities would share a permanent stream, each tapping it high up to draw water to the terraces on either side of it; no mechanism for conflict resolution was needed.

As the demand grows, the supply decreases. The glaciers that plunged deep into the Río Santa canyon have retreated hundreds of meters upslope, and they no longer contribute so much meltwater to the streams. The reduction of water is particularly acute late in the dry season, the time when the farmers plant their early crops and when the valley is still full of tourists. Local government officials and farmers both comment that conflicts over water are more numerous and severe than they were in the past—a trend that seems virtually certain to continue.

Source: Own fieldwork and conversations in March 2009 with Ben Orlove, Columbia University, USA.

When asked about the effects of climate changes on their livelihoods since the start of this century, community members pointed to the following:

- Decrease in profitability of crops.
- Losses in production as a consequence of heavier rains, frosts, droughts, and hailstorms.
- Pests (worms and birds) or diseases affecting potato, onion, and other crops.
- Reduced ability to produce *chuño* (shorter season and fewer viable locations).[10] People elsewhere in the Andes complain of the recent difficulty of preparing *chuño*, which is an important staple because it stores well.
- Loss of wildlife.
- Diseases in livestock resulting from the increase in daytime temperature and decrease in nighttime temperature grades.
- Dry air and strong sun causing sunburn.

Strategies and Adaptations to Climate Change

Facing this situation of uncertainty, some indigenous communities are experimenting with adaptive strategies that are based on new knowledge, including technical advice from agricultural extension agents. At the same time, the indigenous communities are reviving older practices and customs (such as the use of canals in the lakeshore zone) and ancient ceremonies. They also resort to temporary migration. Following are the principal adaptation strategies used by these communities:

- Expanding the cultivation of more profitable new crops, such as irrigated onion, using water pumps and pond construction for water reserves.
- Shifting the potato varieties grown, principally by growing established varieties at higher elevations, but also by experimenting with new varieties.
- Using pesticides for onion and potato crops.
- Changing the time and location for *chuño* production by looking for higher and colder areas.
- Taking advantage of improved climate conditions in higher areas to grow agricultural produce for sale.
- Using traditional practices to fight hailstorms, such as burning pasture to create smoke, even though the community is not always prepared when hailstorms occur.

- Using other communities' practices to fight hailstorms, such as firing off rockets. But this effort has not produced good results because, in the words of one interviewee, "...gunpowder used by our grandparents was different and made the hailstorms run away; now.... it does not work."
- Using techniques learned through training courses, such as protection of crops with covers during hailstorms and improving production with sun tents such as simple plastic greenhouses.
- Reviving the tradition of offering gifts to traditional gods at shrines located in hills, a custom that had otherwise been abandoned.[11]
- Remembering grandparents' advice about traditional indicators to determine sowing dates, including the patterns of stones.
- Accepting the losses and waiting for next year, to start all over again.
- Men going to work temporarily in the cities or the Yungas, getting payment in cash or with products, and women going to buy products in other communities.

People say they are resigned to the situation and do not know what to do next or whom to turn to for help. They fear this is leading to "the end of human life" or, in other words, to the end of a way of life that has sustained these communities for generations.

Potential Indirect Social Impacts

Climate change and variability also have potential indirect social effects in the area. The diminished water supply caused by changes in the hydrological regime is occurring as there is an emerging need for irrigation of traditional crops and for cultivation of new crops that require irrigation. The resulting fall in agricultural output, together with the out-migration of people who used to farm, jeopardizes food security. This leads to more people being affected by under- or malnutrition, which endangers the developmental health of young children and compromises people's immune systems, making them more susceptible to infectious diseases.

The relationship between climatic changes and variability and migration needs further study. In the indigenous communities, young people tend to migrate in search of employment or other income-generating activities. Migration may simply reflect young Aymaras' desire to improve their living conditions by farming in other, more prosperous areas or by securing better-paid jobs in other cities and provinces, but migration appears to be augmented by the effects of climate change and variability. Only a few of the young migrants return to their communities to help relatives at harvest time. Temporary adult migration occurs when the changing climatic conditions cause losses. Temporary migration allows migrants to earn cash, which they use to purchase agrochemicals, tools, and irrigation equipment for their own farming. Whether future extreme climate changes will compel Aymara communities to leave their territories in search of better living conditions should be the object of further research.

A further threat is the loss of traditional knowledge, which jeopardizes adaptation to the climate changes. For example, today there are few farmers who remember and use indicators such as the behavior of birds and insects, which provided their predecessors with guidance for the agricultural calendar.[12]

Certain adaptation measures may turn out to be counterproductive—as in the case of the expansion into new crops such as irrigated onions—and will affect the use of soil and water. The increasing demand and decreasing supply of water will aggravate the squeeze already felt by farmers and will exacerbate the poverty they are already experiencing. This is a phenomenon known all over the Andes and sub-Andes (box 3.3).

The appearance of new pests has been blamed on climate change and variability, though there is some uncertainty whether climatic conditions are indeed the culprit. It may be that the purchase of seeds at local fairs, to introduce new crop varieties inadvertently also introduces pests. Regardless of the cause of these pests, most of the communities are using pesticides to fight the infestations; in the area of study, highly toxic products were being used, polluting the environment and exposing the farmers to severe health risks.

The observed changes in temperatures are likely to have important indirect impacts on health. Rising temperatures mean that malarial mosquitoes will be able to infest areas previously too cold for them to thrive. People living in areas where malaria has not been endemic have no immunity to the disease and will thus be more severely affected than people living in areas with endemic malaria. Indigenous peoples will therefore be at risk from malaria as the habitat of malaria-carrying mosquitoes expands to higher elevations. Indeed, cases of malaria have been registered in a community near Lake Titicaca.[13] It is also possible that the incidence of infectious diseases will increase. Poorer water quality and less water availability may lead to declining standards of hygiene and increases in the incidence of water-borne diseases. And in localities that have recently experienced unusual cold snaps, illnesses caused by the cold have increased.

Box 3.3

Competition and Conflict over Water Resources

The anthropologist Robert Rhoades and his associates have studied social change on slopes of the volcano Cotacachi in Ecuador, one of the Andean peaks whose glaciers have disappeared. His team has shown the dramatic effects of this change on water resources. Though other factors may have also influenced the hydrology of the watersheds that have their highest reaches on the slopes of the mountain, the timing of glacier retreat and disappearance is closely linked to the conversion of many permanent streams to watercourses that are dry for months at a time and to the dropping of the level of a large lake on the mountain.

Competition for these water resources has increased in recent decades. The populations of Quechua farmers high up on the mountain have grown, and commercial agriculture, including export flower production, has increased in lower elevations. These commercial agricultural interests have tapped water sources high on the mountain. In earlier decades, with lower demand, both local and commercial farmers had adequate water supply; now the reduced supplies are insufficient. Incomes have dropped, and many residents complain of the difficulty of meeting even their basic food needs. The level of social conflict has increased dramatically, and the indigenous farmers who were the earlier users of the water often have weaker access to the national judicial system that can address disputes.

In addition, the older generation is deeply troubled that the children will grow up seeing the mountain without its white cover. This white cap is featured in a number of folktales and in songs, and so the local residents sense the disjuncture between their collective memories and the present reality.

Source: Communication in 2009 with Ben Orlove, Columbia University, USA.

Decreases in food production are already causing problems of malnutrition in some communities. Malnutrition can irreversibly damage children's development, especially in those younger than two years. Further, the immune systems of people affected by malnutrition become compromised, leaving them more susceptible to illness.

Sub-Andean Region—Bolivia's Northern Yungas

The parallel mountain range located between the Andes and the plains forms the sub-Andean region. It can be considered as a mountain folded and cut transversely by rivers, with big anticlinal alignments, lengthened asymmetrically with one of the flanks more stretched than others, and presenting a series of hillsides capable of cultivation (Montes de Oca 2004).

The study was performed descending from La Cumbre at 4,660 meters above sea level (Murillo province, La Paz department) down to Coroico at 1,760 meters, passing El Chairo at 1,240 meters (North Yungas province, La Paz department). Areas at different altitudes were included (figure 3.3).

La Cumbre (The Summit)

La Cumbre is situated at 4,660 meters above sea level in the Cordillera Real mountain range. The Real mountain range is the only one that has perpetual snow above 5,300 meters. It is inhabited by Aymara communities devoted mainly to raising camelids and sheep. The area also supports tourism and mineral exploitation activities on a minor scale. Upon descent, some high-elevation crops can be found, such as potato and coca.

Yungas

On the northeastern humid sides of the Andes, the Yungas region forms deep valleys that are located between the Andean high summits and the buttresses of the sub-Andean summits, between 4,200 and 1,000 meters above sea level. The higher zone of this sub-Andean region corresponds

Figure 3.3   Sub-Andean Region of Yungas (meters above sea level)

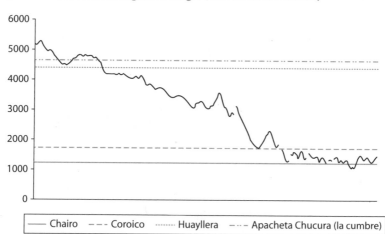

Source: Google Earth and own elaboration.

to "the eyebrow of the mountain," with mixed low woods, fog, and wet *pajonales*[14] (3,400 meters); lower down is a midelevation area of woods with ferns (2,800–2,400 meters); and still lower is the area that corresponds to the real Yungas (lower than 1,500 meters) (Montes de Oca 2004). The region is a center for a great variety of flora, fauna, and endemism in Bolivia.

The climate depends on the altitude, ranging from moderate to subtropical to tropical with a high moisture level. In the higher areas, cloud cover is almost constant, and temperatures are moderate. At lower elevations, there are longer periods of sun and temperatures are higher. The region is exposed to eastern winds that bring rain. The annual average rainfall varies from 900 to 2,500 millimeters.

The population includes some long-established Aymara families of both indigenous and Afro-Bolivian ancestry who are descendants of the workers on haciendas that were expropriated by the 1952 Agrarian Reform. A number of more recent Aymara migrants also live in the study area. Some Quechua and Mosetenes can also be found in other areas of the Yungas.

Indigenous Peoples

Aymara. The settlement of Aymara communities or colonies in the Yungas is very ancient (Preincaic and Incaic periods), and it reflects the efforts of indigenous kingdoms and communities to incorporate this ecological zone and its tropical products, such as coca leaves, hot peppers, and bird feathers, into their economies through so-called vertical ecological integration (Klein 2003). Later, during colonization from the early 16th to the early 19th century, and during the Republic (1828–1899), the Aymara peoples were forcibly moved to work on farms or haciendas to produce food for the city of La Paz and the mining centers. After the Agrarian Reform in 1952, the Aymara peoples were freed and received land titles of deed; from that moment, they were called "Aymara exhacienda." Their descendents still live in the study area. Migrant Aymara peoples can also be found in this area; they settled mainly in the 1980s, finding the living conditions better than those in the Altiplano.

Afro-Bolivian peoples. During the colonial period (end of the 18th century), people were brought from Africa—probably from Senegal, Mozambique, Congo, and Angola—mainly as slaves to work in the Potosí mines. But they could not survive the extreme conditions of altitude and cold weather, so they were taken to work on the haciendas in Chuquisaca and La Paz Yungas. After the Agrarian Reform, the opportunity arose to own land of their own. At present, the Afro-Bolivians are scattered throughout the Coroico district (North Yungas province, La Paz department). They keep their campesino-Aymara-Spanish culture, although they also maintain musical and dancing expressions that are an almost unique Afro-Andean mixture in Latin America (Mihotek 1996). A different view is that these people are historically Aymara who happen to be black, just as there are some Aymara who happen to be white.[15]

Livelihoods
Depending on the elevation, the indigenous communities in the area of study have different livelihood strategies. The Aymara people living in the high area depend on raising llamas, alpacas, and sheep. Cultivation of maize and potatoes for *chuño* occurs only at the highest elevations, which experience hard frosts in the dry season. At intermediate elevations, some arable farming takes place, as well as pig and cattle rearing. In the lower areas, people are mostly devoted to arable farming and trade. Mainly on small plots of land (1–5 hectares), they cultivate tropical crops, including tubers (aracacha, yucca), vegetables, coffee, citrus, other fruit plants, and coca.[16] There is pasture, honey production, use of woodlots for firewood, tourism, mining, and increased colonization.

Trade is important in this region. The productivity of the Altiplano, for all the creativity of its population, has always been limited. Therefore, the highland populations have always practiced vertical ecological integration, interacting with the valley and lowland peoples to obtain basic complementary food items that they could not produce themselves (Klein 2003).

Governance—Transforming Structures and Processes
Structures. Governance is central to the successful support and introduction of adaptation strategies within local communities. During fieldwork for this chapter, community members mentioned a number of organizations and institutions relevant to the support of their adaptation strategies to climate change and variability.

Except in La Cumbre, the forms of traditional community authorities found in the high valleys and the Altiplano are not present among the Aymara and Afro-Bolivian groups of the sub-Andean region, probably because Aymara and Afro-Bolivian groups are not organized at the level of *ayllus* in this region. The sub-Andean groups assumed the model of trade union organization after the Agrarian Reform of 1952, when the haciendas were expropriated. Thus, the most important organizations, as perceived by community members, are the subcentral and central offices of the CSUTCB. Organizations involving economic activity, such as local project-centered organizations in El Chairo and the branch of the national coca producers' association in Coroico, are also important.

Processes. As in the Altiplano, after the Agrarian Reform in 1952 the Aymara and Afro-Bolivian indigenous peoples were freed from the obligation of rendering services on the haciendas, and they received individual titles of deed for the plot of land they had been working on. In this way, the Yungas became an area with family-owned plots.

In this area, two types of migration are in evidence. Short-term and seasonal migration is caused by the need for economic survival; it can be either internal to other communities or areas of the Yungas, or external mainly to the cities of La Paz and El Alto. Permanent migration is mainly of young people who leave to study or work in cities or abroad to improve their living conditions and their social status. They occasionally return during vacations to help their parents during the harvest season or send remittances.

Effects of Climate Change in the Communities Studied

The Aymara and Afro-Bolivian communities of the northern Yungas experience climate change through shifts or fluctuations in the timing of the seasons and increasing irregularity of rainfall, with some intense hailstorms and some periods of drought.

Community members identified the following direct impacts on their livelihoods:

- Decrease in the rainy seasons, hailstorms, and increase in temperature
- Pests
- Water availability in high areas
- For *chuño* production, changes in the length of season and in viable locations
- Decrease in crop profitability
- Losses and decreases in production, particularly of citrus fruits and coffee, because of changes affecting crops, particularly citrus fruit
- Increase in the number of mosquitoes
- Sunburn

Strategies and Adaptations to Climate Change

To face climate change and variability, Aymara and Afro-Bolivian communities have the following strategies and adaptations:

- Shifting the production of *chuño* to higher elevations, where the required colder temperatures can be found
- Traveling further to obtain water for household needs
- Increasing the production of coca to compensate for the loss in productivity of other crops[17]
- Shifting from market agriculture to new income-generating activities, such as tourism, and to higher levels of home consumption of harvests
- Building protective structures along riverbanks to avoid flooding
- Preparing community plans for sustainable development that rest on alternative economic activities such as ecotourism
- Searching for training resources, projects, and institutions that can help deal with climate change and variability without eliminating or decreasing coca crops
- Migrating to other places to obtain wage work, returning to their lands the following year to produce again

Potential Indirect Social Impacts

Climate change and variability have several potential indirect effects in the northern Yungas. As in the Altiplano, an aspect that should be studied

---

Box 3.4

Some Testimonies

"We now live on tourism, we have the hostel and the butterfly shelter, and there are 25 partners."

—Francisca Quispe Flores, Aymara community, 22 years, Chairo, 1,240 meters above sea level

"Those performing only agricultural activities are not doing well, and if they begin changing, transition is not easy, but those who change the activity completely are doing better."

—René Callisaya, Aymara community, 32 years, Chairo, 1,240 meters above sea level

Source: Interviews by the authors in El Chairo (2008).

is the relationship between climate change and variability and migration. Independently of the effects of climate change, the majority of young people in the sub-Andean communities studied tend to migrate to improve their living conditions and social status. They maintain a relationship with their family members remaining in the community, for example, helping them with some productive activities or sending them money and food. Alternatively, the migrants move their families to the places to which they have relocated. This kind of migration produces a loss of cultural identity on the part of young people, who do not wish to be identified in the cities by their indigenous origins.

Young people who stay at home, like some we interviewed in El Chairo, diversify into activities such as tourism and take less and less part in their families' agricultural production, which is used for home consumption only. The result is an erosion of the knowledge pool built up by their ancestors over centuries—knowledge about agricultural practices and about how to read natural indicators that allow weather forecasting and give pointers for the agricultural calendar. This, together with climate change and variability, results in lower agricultural productivity.

In case of losses or lack of production from climate factors, adults migrate temporarily to other areas in the Yungas or to La Paz or El Alto to work as day laborers. It cannot be foreseen with certainty that climate change and variability will force these communities to abandon their homes permanently. However, the possibility of such depopulation seems serious, especially because so many young people have already migrated away. Many do not want their parents to continue farming in their home communities; they rather encourage their parents to join them in their new areas of residence. Or, as 22 year-old Francisca Quispe Flores from the Aymara community of El Chairo said, "The young people have gone to La Paz, and most of the older people who remained as peasant farmers in the village have already died." Migratory movements are also taking place from the Altiplano to the Yungas; because of climatic factors, agricultural workers come to the Yungas in search of better economic opportunities either as producers or as day laborers.

Climate change and variability, the loss of traditional practices and knowledge about production, diversification into other economic activities, and migration endanger the food security of the populations who depend on the production in these areas. Meanwhile, in this traditional coca-producing region, increasing the cultivation of coca to offset the declining yields of other crops will result in an expansion of land under cultivation. Increased soil degradation can be expected, with negative effects on other economic activities such as tourism as the landscape becomes less attractive.

As in the Altiplano, potential human health impacts of climate change include increased incidence of vector-borne diseases such as malaria, as the habitats suitable for malaria-carrying mosquitoes expand. Increased food insecurity will likely lead to more malnutrition, affecting the developmental health of young children and leaving people more susceptible to infectious diseases. Similarly, declining water quality and availability can result in declining standards of hygiene and an increase in the incidence of waterborne diseases. Increased exposure to the sun causes sunburn and increases the risks of eye cataracts.

Summary

Climate change and climatic variability have severe direct and indirect social consequences for the Andean population and the indigenous population in particular. Changes have been observed in the predictability of seasons in the areas studied, altering the agricultural calendar. Extreme climate events such as strong rains, hailstorms, frosts, and mean temperature increases have become more frequent and affect productivity, migration, and people's health. In recent years, people in the areas studied have experienced declining crop yields, crop losses caused by hailstorms, and more frequent frosts that have increased the costs of tilling. The profitability of meat and dairy production has fallen; livestock is losing weight as a result of having to spend more energy in the search for water, while the availability of forage crops is diminishing (MPD/PNCC 2007).

Adaptation strategies are being sought spontaneously in order to survive. These strategies are mainly based on intuition, knowledge transmitted from one generation to another, experience derived from agricultural training, or other experience not related to climate change and variability processes, as well as from imitating successful actions carried out by other communities. In the Andean region, indigenous people's adaptive strategies include the introduction of new crops, the revival of old customs and rituals, and temporary migration. However, people in this region seem resigned to the fact that they lack sufficient knowledge about how to act and about who could help them adapt to these climate changes, which are seen to lead to the termination of the traditional way of life that has sustained indigenous communities for generations. In the sub-Andean region, indigenous people's adaptive strategies include increasing their coca crops, shifting to tourism as a new economic activity, and resorting to temporary migration. Peasant farmers in both regions seek a number of solutions, including modifying their agriculture and migration, but the residents of the sub-Andean region express a greater interest in tourism and in receiving support from NGOs and government programs to address climate change and variability.

People interviewed complained that even if they are economically connected to productive and political organizations, these organizations are not prepared to carry out actions that support them in adapting to climate change and variability. Thus any adaptation efforts are characterized by individual approaches.

In both the high Andes mountain range and the sub-Andean regions, indigenous peoples need to further adjust their livelihood strategies to the changing of water flows. Some have no cultural adaptation strategies for this purpose, and the limited presence of state institutions adds pressure on the capacity of local authorities. Their difficulties in meeting the needs of their community members are seen to lead to increased migration and social change. The pressure from the decreased availability of water that results from climate change, combined with an institutional vacuum, loss of esteem for authorities, and a loss of trust in cultural knowledge, adds threats to traditional culture and religion. There is an urgent need for improved access to information to increase the understanding of the role of knowledge systems for the interpretation of the phenomena and effects of climate change and variability, and to improve the access to resources and relevant institutional capacities.

Notes

1. The Yungas is a stretch of forest along the eastern slope of the Andes Mountains from southeastern Peru through central Bolivia. It is a transitional zone between the Andean highlands and the eastern forests.

2. Between November 2007 and March 2008, for example, La Niña affected 80,000 people in the Bolivian Altiplano, valleys and plains by frost damage, hailstorms, droughts, and floods, causing losses of crops and housing.

3. Alexander and others (2006) and Vincent and others (2005), both with very clear maps, support the claims about thermal amplitude put forth in this chapter.

4. This population is descended partly from the Aymara groups who have lived there since pre-Columbian times (this was an area highly prized for coca production under the Incas), from more recent Aymara migrants from the Altiplano, and from the largely Aymara-speaking Afro-Bolivian communities

founded by slaves brought by the Spanish (communication in 2009 with Ben Orlove, Columbia University, USA).

5. Families have to be registered as members of the community and of the union; it is not enough just to live there. It takes time for people who move in through marriage to be accepted.

6. There is an important link to the Federacion Unica de Trabajadores Campesinas de Bolivia as nearby Achacachi was a major center of Aymara activism that led to the 1952 revolution.

7. The agrarian unions gather in subcentral offices, and the subcentral offices meet at central offices. The majority of the central offices are grouped within a province, but there are also special central offices that do not correspond to the political limits of the provinces. The central offices are grouped in federations. There are 9 departmental federations, 26 regional or special federations, and some national federations; they are all grouped in CSUTCB.

8. An *ayllu* is a grassroots indigenous organization that serves not only social but also political, religious, and ritual functions.

9. Bolivia produces relatively few greenhouse gas emissions, but the country is very vulnerable to the effects of climate change and variability. Various activities have been carried out reflecting the government's commitment to endorse international agreements on climate change and variability. As a result of these activities, an institutional structure in the government has developed a series of tools that define the priorities and lines of action. Priority is given to efforts toward adaptation, with focus on education and investigation. This is reflected in the design and planned implementation of the National Climate Change Adaptation Program.

10. *Chuño* is dehydrated potato, produced in cold, high places.

11. This revival is also a part of the new indigenous movements and does not exclusively reflect climate change (communication in 2009 with Ben Orlove, Columbia University, USA).

12. The use of astronomical observations, such as of the Pleiades, is quite firmly established as a predictor of weather (Orlove, Chiang, and Cane 2000).

13. It seems likely, though, that these people caught malaria in the lowlands; it is difficult to imagine malaria-bearing mosquitoes at such high elevations.

14. A Quechua word for grass-covered highlands.

15. Communication in 2009 with Ben Orlove, Columbia University, USA.

16. This is one of the oldest areas of coca production in the Andes. Most of this coca, chewed in the highlands, goes to the traditional market rather than to the cocaine trade. This is partly because of the nature of the leaf itself, and partly because of the existence of trade links. Located at a higher elevation than newer coca cultivation areas in the Chapare, the coca leaf grows more slowly and has a lower cocaine content but a better flavor.

17. Naturally, the fluctuations in coca production also reflect changes in market conditions and in government policy.

References

Alexander, L. V., X. Zhang, T. C. Peterson, J. Caesar, B. Gleason, A. M. G. Klein Tank, M. Haylock, D. Collins, B. Trewin, F. Rahimzadeh, A. Tagipour, P. Ambenjek, K. Rupa Kumar, J. Revadekar, G. Griffiths, L. Vincent, D. Stephenson, J. Burn, E. Aguilard, M. Brunet, M. Taylor, M. New, P. Zhai, M. Rusticucci, and J. L. Vazquez-Aguirre. 2006. "Global Observed Changes in Daily Climate Extremes of Temperature and Precipitation." *Journal of Geophysical Research* 111 (D5).

CIDOB (Confederation of Indigenous Peoples of Bolivia). 2007. "Major Indigenous Peoples in Bolivia." http://www.coica.org.ec/ingles/members/cidob.html#.

Georges, Christian. 2004. "20th-Century Glacier Fluctuations in the Tropical Cordillera Blanca, Peru." *Arctic, Antarctic, and Alpine Research* 36: 100–107.

Gil, Vladimir. 2008. *Adaptation Strategies to Climate Change: A Case Study of the Societal Impacts of Tropical Andean Glacier Retreat.* New York: The Earth Institute, Columbia University.

Google Earth. 2008. http://earth.google.com.

Instituto Nacional de Estadística (INEC). 2001. VI Population Census.

IPCC (Intergovernmental Panel on Climate Change). 2007a. *Climate Change 2007: The Physical Science Basis.* Contribution of Working Group I to the *Fourth Assessment Report* of the IPCC. Geneva: IPCC. http://www.ipcc.ch.

———. 2007b. *Climate Change 2007: Impacts, Adaptation, and Vulnerability.* Contribution of Working Group II to the *Fourth Assessment Report* of the IPCC. Geneva: IPCC. http://www.ipcc.ch.

IWGIA (International Work Group for Indigenous Affairs). 2008. *The Indigenous World Yearbook.* Copenhagen: IWGIA.

Kaser, G. 1999. "A Review of the Modern Fluctuations of Tropical Glaciers." *Global and Planetary Change* 22: 93–103.

Klein, Herbert S. 2003. *A Concise History of Bolivia.* Cambridge, U.K.: Cambridge University Press.

Layton, Heather Marie, and H. A. Patrinos. 2006. "Estimating the Number of Indigenous People in Latin America." In *Indigenous Peoples, Poverty, and Human Development in Latin America*, eds. Gillette Hall and H. A. Patrinos. New York: Palgrave.

Mihotek, K., ed. 1996. "Comunidades, territorios indígenas y biodiversidad en Bolivia." Universidad Autónoma Gabriel René Moreno, Centro de Investigación y Manjeo de Recursos Naturales Renovables, Santa Cruz, Bolivia.

Montes de Oca, I. 2004. *Enciclopedia geográfica de Bolivia.* La Paz, Bolivia: Atenea S.R.L.

MPD/PNCC (Ministry of Planning and Development, National Programme for Climate Change). 2007. *Annual Report.*

Orlove, Ben. 2009. "The Past, the Present and Some Possible Futures of Adaptation." In *Adapting to Climate Change: Thresholds, Values, Governance,* eds. W. Neil Adger, Irene Lorenzoni, and Karen O'Brien, 131–163. Cambridge, U.K.: Cambridge University Press.

Orlove, B. S., J. C. H. Chiang, and M. A. Cane. 2000. "Forecasting Andean Rainfall and Crop Yield from the Influence of El Niño Pleiades Visibility." *Nature* 403: 68–71.

Salick, Jan, and Anja Byg. 2007. *Indigenous Peoples and Climate Change.* Oxford, U.K.: Tyndall Centre for Climate Change Research for Climate Change Research.

Verner, Dorte. 2010. *Reducing Poverty, Protecting Livelihoods and Building Assets in a Changing Climate: Social Implications of Climate Change in Latin America and the Caribbean.* Washington, DC: World Bank.

Vincent, L. A., T. C. Peterson, V. R. Barros, M. B. Marino, M. Rusticucci, G. Carrasco, E. Ramírez, L. M. Alves, T. Ambrizzi, M. A.Berlato, A. M. Grimm, J. A. Marengo, L. Molion, D. F. Moncunill, E. Rebello, Y. M. T. Anunciacão, J. Quintana, J. L. Santos, J. Baez, G. Coronel, J. García, I. Trebejo, M. Bidegain, M. R. Haylock, and D. Karoly. 2005. "Observed Trends in Indices of Daily Temperature Extremes in South America, 1960–2000." *Journal of Climate* 18 (23): 5011–5023.

CHAPTER 4

Indigenous Peoples of the Caribbean and Central America

The Central American and Caribbean subregion has become widely known for the increased intensity and frequency of natural disasters. It also harbors some of the world's most affected and vulnerable populations with regard to the effects of climate change and variability.

The most significant sources of impact from climate change and variability on this subregion are a continuing upward trend for intense precipitation and extended periods of drought resulting from higher temperatures. Hurricanes, too, can have a devastating impact and are expected to increase in intensity.[1]

Although extreme events like hurricanes receive unprecedented coverage on global media networks, and the related dramatic social impacts are often obvious and on many people's minds when the subject of climate change and variability arises, the slower-onset effects of climate change—such as altered precipitation patterns, unpredictable appearances of frost and hail, and prolonged periods of dry days—can have just as serious effects on local populations, their assets and access to resources, and their transforming structures and processes. These changes influence livelihood strategies and livelihood outcomes as well as the vulnerability context within which people live. This chapter puts particular emphasis on indigenous people's assets. Although not following it strictly, this chapter is inspired by the analytical framework developed by sociologists such as Bourdieu (1986) and Bebbington (1999), and the Sustainable Livelihoods Framework promoted by the United Kingdom's Department for International Development (DFID 2001).

Although indigenous peoples from the Carib, Chibcha, Maya, Aztec, Arawak, and other civilizations have dispersed across the entire Caribbean and Central American isthmus, the largest concentration today, approximately 20 million indigenous people, is found on the mainland from Nicaragua to Mexico. table 4.1 and box 4.1 present an overview.

Abrupt Effects of Extreme Events

Coastal Communities in Nicaragua

Large numbers of indigenous people in Nicaragua live on the Caribbean coast, which is highly susceptible to hurricanes and storms. This area was selected as a case study for both Nicaragua and the larger region of the Caribbean and Central America to discuss the impact of intensified storms and hurricanes on indigenous peoples' well-being and livelihoods. Hurricanes directly affect coastal regions and can create flooding in low-lying, poorly drained areas. Figure 4.1 shows all hurricane occurrences in the case study area since 1911, and it also shows the location of the indigenous communities in the area and the boundaries of the indigenous territories by ethnic group.

Two field sites on the Nicaraguan Caribbean coast were chosen for this case study: a multiethnic territory (made up of Miskitu, Creole, and Garifuna communities) located in the Autonomous Region of the South Atlantic (RAAS), which was affected by Hurricane Joan in 1988, and a

Table 4.1   Facts on Indigenous Peoples in the Central American Isthmus

Country	Number of peoples	Approximate number of indigenous individuals (1000)	Percent of total population
Belize[a,b]	2	55	17–20.0
El Salvador[a,b]	3	400	8.0
Guatemala[a]	20[d]	6,000	60.0
Honduras[b,d]	7	75–500	1–7.0
Mexico[a,c]	61	12,400	13.0
Nicaragua[a]	7	300	10.0
Panama[a]	7	250	8.4

Sources: Authors' elaboration based on following sources: (a) by self-identification (IWGIA 2008); (b) Layton and Patrinos 2006; (c) Comisión de Desarrollo Indígena (CDI); and (d) U.S. Central Intelligence Agency World Fact Book.

Box 4.1

Indigenous Peoples in Mexico and Nicaragua

Mexico

The indigenous population of Mexico is estimated at 12.4 million, or 13 percent of the country's total population, spread across the country's 32 states. Sixty-eight indigenous languages were listed in 2008, spoken in 368 variants grouped in 11 linguistic families.

Mexico ratified ILO Convention 169 in 1990, and in 1992 Mexico was recognized as a pluricultural nation when Article 4 of the Mexican Constitution was amended. In 1994, the Zapatista National Liberation Army (*Ejército Zapatista de Liberación Nacional*, EZLN) took up arms in response to the misery and exclusion suffered by indigenous peoples. The San Andrés Accords were signed in 1996, but it was not until 2001 that Congress approved the Law on Indigenous Rights and Culture—and even that law did not reflect the territorial rights and political representation enshrined in the San Andrés Accords. More than 300 challenges to the law have been rejected. From 2003 onward, the EZLN and the Indigenous National Congress (*Congreso Nacional Indígena*) began to implement the San Andrés Accords throughout their territories, creating autonomous indigenous governments in Chiapas, Michoacán, and Oaxaca. Although the states of Chihuahua, Nayarit, Oaxaca, Quintana Roo, and San Luís Potosí have state constitutions with regard to indigenous peoples, indigenous legal systems are still not fully recognized.

Nicaragua

The seven indigenous peoples of Nicaragua live in two main regions: the Pacific Coast and Center North of the country (or simply the Pacific), which is home to four indigenous peoples—the Chorotega (82,000), the Cacaopera or Matagalpa (97,500), the Ocanxiu (40,500), and the Nahoa or Náhuatl (19,000)—and the Caribbean (or Atlantic) Coast—where the Miskitu (150,000), the Sumu-Mayangna (27,000) and the Rama (2,000) live. Other peoples enjoying collective rights in accordance with the Political Constitution of Nicaragua (1987) are those of African descent, who are known in national legislation by the name of ethnic communities. These include the Kriol or Afro-Caribbeans (43,000) and the Garifuna (2,000).

Only in recent years have initiatives been taken to improve regional autonomy. Regulations for this purpose include the 1993 Languages Law; the 2003 General Health Law, which requires respect for community health models; Law 445 on the

(continued)

Box 4.1 *(continued)*

System of Communal Ownership of Indigenous Peoples and Ethnic Communities of the Autonomous Regions of the Atlantic Coast of Nicaragua and of the Bocay, Coco, and Indio Maíz river basins, which came into force at the start of 2003; and the 2006 General Education Law, which recognizes a Regional Autonomous Education System (Sistema Educativo Autonómico Regional).

In the first section of this chapter we investigate the rapid-onset impacts of extreme climate-related events through fieldwork on the Caribbean coast. Nicaragua, and in particular its Caribbean coast, serves as a case of the highly vulnerable areas within the region of Central America and the Gulf of Mexico that have been continuously hit by extreme events; our findings from Nicaragua are supported by studies from primarily the Mexican state of Chiapas. Slow-onset climate-change phenomena, too, are very much present in the Caribbean and the Central American isthmus. In the second section of the chapter we explore these changes using data from field visits to indigenous peoples' coffee production sites in Chiapas—supported by information from Nicaragua and Peru—and from places of indigenous peoples' maize, avocado, and forest production in Michoacán, Mexico. Further studies in, for example, small island states of the Caribbean may have added to the insights of how extreme events are perceived among local populations; however, given our focus on indigenous populations we find the geographical focus on Mexico and Nicaragua is justified. The third section of the chapter looks at the potential indirect social impacts of climate change and variability, and the fourth section concludes.

The Sandinista National Liberation Front (*Frente Sandinista de Liberación Nacional,* FSLN) came to power in Nicaragua in 1979, soon after having faced an armed force supported by the United States. The indigenous peoples of the Caribbean coast, particularly the Miskitu, formed a part of this force. To put an end to indigenous resistance, the FSLN created the Autonomous Regions of the North and South Atlantic (RAAN and RAAS) in 1987, on the basis of a New Political Constitution and Autonomy Law 28). Three years later, the FSLN lost the national democratic elections in Nicaragua to the Constituent Liberal Party (*Partido Liberal Constituyente*), and an agricultural policy was implemented that promoted colonization and individual titling of indigenous territories, also commencing the establishment of protected areas over these territories, without consultation.

Source: IWGIA 2008.

Figure 4.1   Climate Change and Indigenous Communities on the Nicaraguan Atlantic Coast

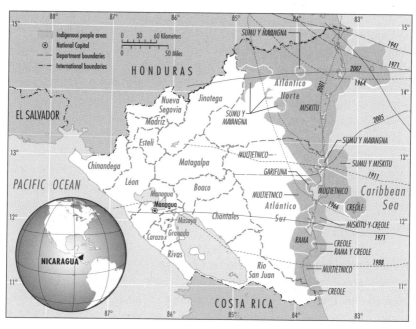

IBRD 37805
MAY 2010

Source: Adapted based on own data partly developed for UNDP, Nicaragua (2005).

group of Miskitu communities in the Autonomous Region of the North Atlantic (RAAN) affected by Hurricane Felix in 2007. The two field sites were selected to enable a comparison of the experiences and perceptions of the social impacts of hurricanes over time by local indigenous groups, obtained through a maximum variation sample.

Together, RAAN and RAAS have a population of around 500,000 (roughly 10 percent of Nicaragua's total population), who are sparsely distributed in small communities of 200–3,000 people, mostly in RAAN. The main indigenous groups present are the Miskitu, the Sumu-Mayangna complex (Ulwas, Twaskas, and Panamaskas), and the Rama, all of whom are linguistically related and belong to the Macro-Chibcha linguistic family. Other ethnic communities include Creoles and Garifuna (both Afro-descendants). The former speak a variation of the Creole English that is common in the Caribbean Islands, and the latter a language belonging to the Arawakan family.

Inland Forest Communities in Mexico—Chiapas

Although Mexico's southern region is known for geological disturbances, hydro-meteorological natural disasters have become increasingly common and cause high economic losses. Of the losses shown in table 4.2 and figure 4.2, 68 percent derived from hurricanes between 1985 and 2005. While Hurricanes Wilma and Stan both hit southern Mexico causing high economic damage, the adaptive capacity of the affected populations and hence the social impacts of the two storms differed substantially. Hurricane Wilma destroyed tourist facilities on the Yucatan peninsula, which were to a large extent insured. Hurricane Stan hit the uninsured assets of the local and largely indigenous population living in urban slums and as subsistence farmers in Chiapas regions such as Escuintla, Mapastepec, and Cacahoatán. Some relief was forthcoming from the Mexican government and NGOs, and local people could mobilize support from relatives who had migrated, but Hurricane Stan hit hard and destroyed many assets.

Direct Social Impacts from Climate Change

It has been said that the social impacts of climate change and variability are increased by the alterations these trends are causing in the natural environment. This is particularly true for indigenous people in these parts of Mexico and Nicaragua. They often rely heavily on natural resources directly for food security and health, and also indirectly for their livelihood and well-being. In fact, their cultural institutions, knowledge, and practices have evolved around the use and conservation of particular natural resources of the region they live in (Balée 1989; Kronik 2010). However, their production systems are dynamic; for example, they have incorporated coffee

Table 4.2   Major Hurricanes in Mexico since 1982

Year	Hurricane	Economic loss (US$ million)
1982	Paul	82
1988	Gilbert	597
1990	Diana	91
1993	Gert	114
1995	Opal	151
1997	Paulina	448
2002	Isidora	235
2005	Emily	845
2005	Wilma	1,788
2005	Stan	2,006

Sources: Conde Àlvarez and Saldaña Zorrilla 2008; CEPAL 2006; Carpenter 2006; SEGOB 2002.

Figure 4.2   Frequency and Economic Losses from Natural Disasters in Mexico, 1970–2000

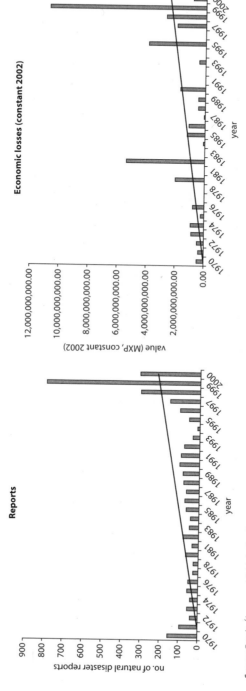

Source: Conde Álvarez and Saldaña Zorrilla 2008.

well—a crop that fits better into small-scale, household agriculture than sugarcane or cotton.

Figure 4.3 and figure 4.4 show where productive systems of the communities at the two Nicaraguan field sites were affected by Hurricanes Joan and Felix.

Even though most of the indigenous communities of the area have a combined subsistence livelihood (mostly a mixture of fishing, agriculture, and hunting), the Miskitu occupy all the productive zones in the area (both in RAAN and RAAS), putting them in a better position to adapt their livelihood strategies in the event of hurricanes and storms (table 4.3). Also, other communities (mostly coastal) are slowly inserting themselves into an incipient market economy with activities that are linked to national or international markets (shrimp, lobster, timber, artisanal gold mining), giving them access to building materials and other goods that make them better prepared to withstand climate change and variability events.

Mestizo people are found on both extremes of the social spectrum, from rich cattle ranchers and traders to very poor subsistence producers of corn and beans. As discussed further below, one of the unexpected findings of this study was that mestizo subsistence farmers are even more vulnerable to hurricanes than are indigenous people because of their physical isolation, lack of assets, and cultural differences that pose other obstacles to the delivery of aid (table 4.3).

Figure 4.3   Areas Affected by Hurricane Joan, 1988

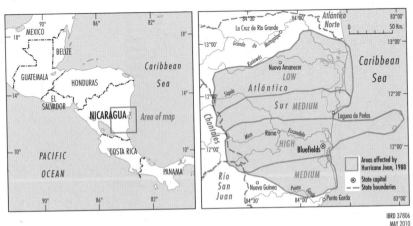

IBRD 37806
MAY 2010

Sources: Authors' elaboration based on data from the U.S. Department of Commerce (2009) National Climate Data Centers (NOAA), and Nitlapan, Universidad Centroamericana (UCA).

Figure 4.4   Areas Affected by Hurricane Felix, 2007

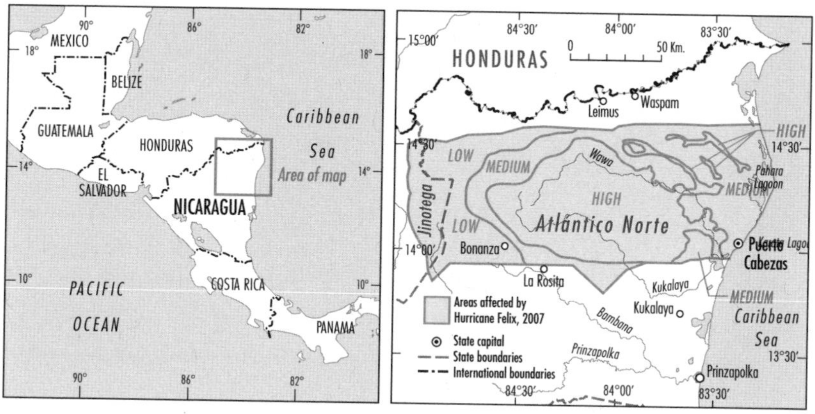

IBRD 37807
MAY 2010

Sources: Authors' elaboration based on data from the U.S. Department of Commerce (2009) National Climate Data Center NOAA, and Nitlapan, UCA.

Table 4.3   Description and Characteristics of Each Productive Zone Shown in Figures 4.3 and 4.4

Indicator	Coastal-marine (fisheries)	Pinewood savannah	Rainforest and mining mangroves	Agricultural zones (different subzones)
Population density (no. of people/km²)	1 to 3	1 to 3	1 to 5	10 to 34
Ethnic group	Mískito Creole Garifuna Rama	Mískito	Mayangnas Mískito Mestizo	Mayangnas Mískito Mestizo
Main productive activity (livelihood)	Lobster Shrimp Scale fish Turtle Subsistence agriculture (rice, cassava, plantain, banana, fruit, etc.)	Timber extraction (pinewood) Charcoal and firewood nontimber forest products (e.g., fruits)	Timber Gold and silver extraction (commercial and artisanal)	Cattle ranching (milk products and meat) Basic grains (rice and beans) Export crops (cocoa, cassava, etc.) Hunting

(continued)

Table 4.3   Description and Characteristics of Each Productive Zone shown in Figures 4.3 and 4.4 *(continued)*

Indicator	Coastal-marine (fisheries)	Pinewood savannah	Rainforest and mining mangroves	Agricultural zones (different subzones)
Main market	USA (Miami) Local and national market	Local and interregional markets (RAAN-RAAS)	Lumber: Caribbean islands and national markets Gold and silver: national market and Canada	USA (Miami) Caribbean Islands El Salvador Costa Rica National and local markets

Source: Based on own data obtained from fieldwork in the region (2008).

Most people interviewed in Nicaragua and Mexico agreed that the main social impact of hurricanes on their livelihoods is access to and availability of crops, forest, and fish resources. People affected by Hurricane Joan, 20 years ago, mentioned that they had yet to see a full recovery of the abundant forest resources that existed before the hurricane (especially lumber and wildlife) (box 4.2). In contrast, however, they noted that fish catches and agricultural yields had returned to close to their prehurricane levels; recovery took five years for fishing and perennial crops such as fruit trees, and just one year for rice, beans, banana, plantain, cassava, and maize.

According to the interviews, the main reason why the forest resources have not fully recovered is the advance of the agricultural frontier. This finding merits further research. It is said that after hurricanes, poor mestizo farmers are better able to invade indigenous territories, setting fire to trees knocked down by the intense winds and claiming more land for basic-grain agriculture and cattle ranching. Many times, their fires get out of control, affecting larger areas than intended, destroying the seed banks from which forests regenerate naturally and delaying full recovery by many years (Bradford 2002).

According to the interviewees on the Nicaraguan Caribbean coast, indigenous people in this part of the world are very attached to their communities and territories. Here we can draw a parallel to the Peruvian case near the glaciers (described in chapter 3), where individuals migrate but much prefer to stay within their territories when they do so, and also count on at least some family member remaining in the community of

Box 4.2

Edited Testimony of April 2008 by Oswaldo Morales, Garifuna, Member of the National Commission for Territorial Demarcation of Indigenous Land, Laguna de Perlas, Nicaragua

Livelihood, Resources, Institutions

The resources tend to be scarcer today than before Hurricane Joan even after 20 years of recovery, especially forest resources and wildlife. Nevertheless, people combine different activities, mainly for subsistence and small-scale production for income generation, ranging from agriculture, forest products extraction, and sand mining to fisheries.

People in the poorest community in the area concentrate on farming and subsistence fishing; before Hurricane Joan they did more hunting. Few do commercial fishing. The most vulnerable community in terms of location was flooded during the hurricane, and most of the houses were destroyed. It received more aid than other communities to rebuild housing.

A new road connecting the Pearl Lagoon area to the rest of the country also gives more access to farmlands by vehicles. This has created a modest tendency for community members to go back to farming.

Indigenous communities are tending to rely more on fishing, instead of a combination of fishing and agriculture, due to market demand and the war of the 1980s. At the same time, poor mestizo farmers began the invasion of community territories in search of agricultural lands. These farmers increased their access to "available" lands after the passage of Hurricane Joan.

In the 1980s (during Joan) there were fewer organizations than today working in the field of disaster management and community development in general. Central government and religious organizations were the main sources of aid after Hurricane Joan. Nowadays governmental and nongovernmental actors provide aid, including the autonomous regional government, municipal government, rural development NGOs, and an integrated disaster management committee, where all these actors interact and plan actions. Today people are well informed about hurricanes and the different types of organizations working in the field of hurricane mitigation.

Adaptation Strategies

The dwindling of natural resources and the lack of job opportunities in the region move some members of the community to look for temporary jobs on cruise

(continued)

Box 4.2 *(continued)*

ships on the Caribbean. Others move to the highland savannah to avoid flooding and direct wind.

Hurricane Joan left people more aware about hurricanes and the need to be prepared for such events within the limits of their capacity. People are better informed than before as they are provided with regular information through a community radio station about possible hurricanes and storms.

Now, people trust the hurricane forecasts. Learning from previous experience, people are more willing to temporarily evacuate their homes. Also wooden houses are being replaced with cement ones.

origin. Migration after events such as hurricanes tends to be a temporary strategy, and people mostly return to their community as soon as basic conditions are reestablished. People mention that in the past, they (especially older people) were very reluctant to abandon their communities, and this resulted in an increased number of fatalities, but that now—having seen the damage caused by hurricanes and been influenced by municipal campaigns—they do not hesitate to evacuate their homes and communities if they feel they are not safe. Even though these communities are considered poor, their members generally do not stay away for long. One of the main explanations given is the rights to and control of a relative abundance of natural resources such as lobster and shrimp, which maintain high prices on the international market. However, the dwindling of natural resources, compounded by climate change and variability and the lack of job opportunities in the region, is making younger people search for temporary work on cruise ships in the Caribbean. The inclination to learn English (Creole English) by indigenous youth is clearly seen as part of a strategy to expand their livelihood options. This affects cultural identity by creating tensions between traditions and modernity, between mother tongue and the *lingua franca*.

Since vulnerability is considered a function of economic, social, cultural, political, environmental, and technological assets, table 4.4 summarizes the livelihood assets and sources of vulnerability to climate change and variability that impact indigenous groups and poor mestizo communities in the Nicaraguan areas studied.

Based on his experience as member of an evacuation and emergency team during Hurricane Joan in 1988, one of the people interviewed

Table 4.4 Typical Assets Held by Indigenous Coastal and Poor Inland Mestizo Communities

Asset endowment	Indigenous coastal communities	Poor inland mestizo communities
Material assets	Slowly, concrete walls and metal roofing are being introduced after the experiences with hurricanes, changing the aspects of many communities and providing more secure housing to withstand hurricanes. Access to expanded markets for fishing products is the main source of stable income.	Housing remains traditional: lumber with thatch roof. Income, mainly from charcoal and basic-grains production, remains low.
Informational assets	With the introduction of electricity to many communities, radio and TV audiences tend to increase, which gives them access to information on hurricane early warnings.	Radio access is relatively good with the introduction of community radios. However, difficult physical access to these isolated communities makes them vulnerable in terms of evacuating people to safe shelters.
Natural assets	Before, subsistence agriculture and fishing were equally important for livelihoods. In the 1990s, a commercial artisanal fishing tendency started, which continues until today. However, with the opening of a new road, coastal people are going back to farming with the hope of new markets for agricultural products. Access to good quality water tends to be a problem, which worsens temporarily after a hurricane. Many aid organizations help with rehabilitation of wells.	Tend to focus on agriculture and cattle ranching. The road presents a good opportunity for them. Tensions with indigenous peoples continue over access to land and forest resources. Access to good quality water is usually a problem, which worsens after a hurricane. Sources of drinking water are creeks and rivers (which are usually polluted with coliforms), not water wells.
Financial assets	Few community members have savings or access to credit that will facilitate adaptation measures; those who have credit or savings are mostly traders, shop owners, and fisher-people.	Because of isolation of these communities, almost no one has bank savings or access to credit that will facilitate adaptation measures.
Human assets/ psychological assets	Primary and secondary education is slowly being introduced in indigenous communities. Previous experiences with hurricanes give them the knowledge and awareness of the effects of hurricanes and how to prepare better for such events.	Illiteracy remains high. Newcomers to the agricultural frontier have little knowledge of hurricane effects.

(continued)

Table 4.4 Typical Assets Held by Indigenous Coastal and Poor Inland Mestizo Communities *(continued)*

Asset endowment	Indigenous coastal communities	Poor inland mestizo communities
Social assets/ organizational assets	Central government and religious organizations were the main actors to provide aid after Hurricane Joan. Different governmental and nongovernmental actors—autonomous governmental and nongovernmental actors—autonomous regional governments, municipal governments, rural development NGOs—were at work during and after Hurricane Felix. Also, the municipalities of the area have emergency plans and are organized in committees for evacuation and food delivery to victims of hurricanes. These committees receive regular training for the National System for Disaster Prevention.	Community organizations exist, but access to municipal and central government organizations remains weak.
Cultural assets	Legal advances have been made, such as the Nicaraguan Autonomy Law for the two Atlantic regions, home of most Nicaraguan indigenous people. Nicaragua has an autonomy law that gives indigenous people the right to self-government and the usufruct of the natural resources found in the area based on their traditions and historical rights. The law also gives these communities the right to receive education in their native languages. This gives them a kind of "coast people identity," which unifies them before the national state and the dominant mestizo culture on issues related to development and political participation in the national arena.	The fact that this group has cultural and language affinity with the dominant culture of the country gives them an advantage when it comes to access to the national markets, not only for the basic grains they produce, but also for selling large tracts of cleared land to rich cattle ranchers and traders from the interior.

Source: Based on own data collection.

mentioned three factors influencing a community's vulnerability and self-rehabilitation capacity (adaptation) during and after hurricanes:

- *Location* – communities closer to the beach line and at sea level are more vulnerable, even if they have adequate buildings.
- *Disposition* – smaller and single-ethnic communities show more social cohesion than multiethnic or larger communities; therefore, the former show more disposition to collaborate or help each other after a climate event without relying solely on external aid.[2]
- *Quality of infrastructure and buildings* – communities with more robust houses tend to better withstand storms and hurricanes; however, if such communities are at the ocean shore, they are still highly vulnerable to such climate events.

Table 4.5 summarizes the main findings on the social impact of intensified hurricanes and storms in the Nicaraguan study areas.

Effects of Slow-Onset Climate Change Processes

The slow-onset processes associated with projected climate change will have consequences that are at least as severe as those of hurricanes, cyclones, and strong storms. Thus increased vulnerability can be expected for large populations, not least for the indigenous populations, with very similar impacts to those suffered in the Andes, sub-Andes, and the Amazon regions.

The discussion below focuses on the impacts of slow-onset hazards occurring and projected for the Caribbean and Central America: higher mean temperatures with increasing numbers of dry days, unpredictable precipitation (including torrential rain and hail), and frosts. Indigenous communities in Mexico depend on agriculture for their livelihood. In Mexico, the dry-day frequency trend continues upward, and robust seasonal signals point toward an overall reduction in precipitation in June, July, and August. During these months, people have counted on the *canícula*, or break in the rainy season, that is associated with the movement of the Intertropical Convergence Zone (ITCZ). If the *canícula* gets longer, it threatens crops, especially maize. In December, January, and February, the drying trend is very robust and strongest in northern Mexico.

In the Mexican states of Michoacán and Veracruz, there are many poor nonindigenous/mestizo populations interspersed with indigenous

Table 4.5 Nicaragua: The Social Impact of Intensified Storms and Hurricanes on Indigenous Coastal and Poor Inland Mestizo Communities

Environmental conditions, including land, forest, and water resources	Hurricanes heavily affect crops, water wells (with contaminated rainwater), and forests. Based on experiences from Hurricane Joan and the impact of Hurricane Felix, the time needed to recover for water and annual crops is rather short; fishing resources and perennial crops take a few years (3–5 years), and for forest resources, the estimate is between 10 years (mangrove forests) and 50 years (rain forests).
Livelihoods – poor agrarian societies in dry-land ecosystems	Not applicable for the case of Nicaraguan Atlantic coast; however, in the case of Hurricane Mitch (1998), inland communities of the central dry-land ecosystems of Nicaragua and Honduras were severely affected. The local livelihoods were principally affected by landslides from the intense precipitation (four years of rain fell in four days). In this case, the direct social impacts include increased poverty levels and poor health conditions (leptospirosis) because of heavily degraded assets caused by soil erosion and flooding in already marginal land; limited access to administrative centers because of road and other infrastructure destruction, which affected health conditions and other emergencies; and lack of institutional preparedness to respond to the event, which retarded aid reach.
Livelihoods – low-income communities in coastal areas	These communities tend to be fishing communities or they may combine fishing activities with agriculture and forest resource extraction (timber and nontimber). Therefore, the main impact would be on the access to and availability of these resources. In the case of Hurricane Joan 20 years ago, people mentioned that they have not fully recovered when it comes to forest and wildlife resources, while fishing activities and agriculture tended to be close to levels previous to the event after one to five years.
Livelihoods – low-income communities in urban areas	Destruction of economic infrastructure: Bluefields, Hurricane Joan (1988); Puerto Cabezas, Hurricane Felix (2007), are the main towns of these two indigenous peoples' regions. Poor neighborhoods, where indigenous people are concentrated, usually do not have insurance, savings, or access to credit to facilitate adaptation measures.
Health	Impact on good water quality affects mainly children and older people with diarrhea and new, potentially deadly diseases, including leptospirosis, which is transmitted by rat urine. Rats in human dwellings proliferate during and after flooding.

Conflicts regarding access to assets, including land tenure	Main conflicts exist between mestizo and indigenous groups over access to farmland and forest resources with the advance of the agricultural frontier. This tends to move faster after the passage of hurricanes and the subsequent forest fires at the start of the next planting season.
Migration	According to the perception of the interviewees, indigenous peoples in this part of the world tend to be very attached to their communities and territory; thus, migration after events such as hurricanes tends to be temporary, and people tend to return to their communities as soon as the basic conditions are reestablished. In some cases, people are very reluctant to abandon their homes, resulting in an increased number of fatalities. Even though these communities are considered poor, their members generally refrain from migrating for extended periods, still having quite abundant natural resources like lobster and shrimp, which maintain high prices on the international market. There is a tendency for younger people to get temporary jobs on cruise ships in the Caribbean.
Instruments to support the mitigation of climate change	A national system for disaster prevention and mitigation is in place (actors include the army, police, Ministries of Health and Transport), which coordinates with indigenous peoples' municipal and local leaders to supply aid (food, water, clothing, medical attention); evacuate people to appropriate shelters; and provide people with building materials for housing, well rehabilitation, and seeds for the next plating season after the hurricanes.
Instruments to support the adaptation to climate change	With regard to formal instruments, early warning mechanisms for hurricanes and floods are in place. In the case of informal instruments, natural resources tend to be scarcer today than before hurricanes, in some cases even after 20 years of recovery (Hurricane Joan), especially forest resources and wildlife. Thus, as part of their livelihood adaptation, people are combining different productive activities, from agriculture, forest products extraction, and sand mining with fisheries and seasonal work on cruise ships. These activities are mainly for subsistence and small-income generation, but they are not sufficient for people to capitalize and acquire complex means of production (except for fisheries and cruise ship jobs).

Source: Based on own findings from interviews and secondary sources.

groups, and they suffer the same vulnerabilities. In Chiapas, by contrast, there are large areas that are exclusively indigenous. In Michoacán, indigenous people cultivate maize, typically in association with beans in much of the highland, and complement this to different degrees with fruit and forest production. In the states of Chiapas and Veracruz, coffee production is more prominent. Various recent studies show that although these crops depend upon different conditions (including precipitation and temperature), they are all likely to be affected by climate change and variability.

Chiapas

Chiapas is one of the Mexican states richest in biodiversity. Its climatic conditions are highly complex because of its undulating topography. Situated in the south of the country, bordering Guatemala and the Pacific coast, Chiapas is Mexico's main coffee-producing state, with an average annual production that surpasses 115 million kg. One million jobs depend on coffee production in the state, and 60 percent of the producers are smallholders farming less than five hectares each. Large indigenous populations, such as the Tzotzil and Tzeltal, live at higher elevations that are too cool for coffee, but many of these people migrate, like poor indigenous people from Guatemala, to work on the coffee harvest.

The majority of the smallholder coffee growers are indigenous people living in areas marginalized by state and federal institutions and with limited access to markets. Their traditional means of production, with forest shade cover and high biodiversity, meet the requirements of niche consumers demanding coffee that is organic and bird-friendly (SAGARPA 2008). The approximate area covered by organic coffee is 35,000 hectares (Conde Álvarez and Saldaña Zorrilla 2008).

Both slow- and rapid-onset climatic changes and variability affect coffee production and, hence, the livelihoods depending on it. Adaptation for Smallholders to Climate Change (AdapCC)[3] is an initiative developed to identify the vulnerability of the crop to climate change and variability and enhance smallholders' capacities to deal with the negative impacts. This three-year private-public partnership involves Peruvian, Nicaraguan, and Mexican organizations of cooperatives along with British coffee importer Cafédirect PLC and the German technical cooperation agency (GTZ). Researchers have run climate change and variability scenarios for the area that project temperature increases of from 1°C to 2°C by 2025 and from 1°C to more than 8°C by 2080. Most scenarios project decreased precipitation. Smallholders have already lost adaptive capacity as the result of soil degradation, lack of water, and natural hazards (frosts, torrential rains on high-altitude hillsides, and drought at lower altitudes). Low productivity and prices seem to be a primary development obstacle. AdapCC predicts serious threats and consequences for coffee-producing families. Productivity will decrease dramatically and production may even disappear in the lowlands, resulting in migration pressures toward higher altitudes and into the jungle areas, putting pressures on the Lacandon Maya peoples.[4] As the area suitable for coffee production diminishes, production will be concentrated in a smaller area. World market prices may become even more volatile, and incomes and exports are likely to decline. Plant diseases will likely increase as a result of the projected changes in precipitation and temperature, affecting crop development and decreasing yields (Alfaro, Saladaña, and Linne 2008).

Just as unexpected rains destroy fruit production in the orchards of the lower Andes and Amazon, similar effects may occur in coffee production in the Central American isthmus, as shown by AdapCC.[5] Coffee production depends on certain climatic conditions that may not persist much longer.[6] Unexpected rains in April, as seen in recent years, can destroy coffee flowers and severely affect production. Adding to the adversity, the summer dry period, *canícula*, is likely to last longer. This may affect agriculture, with increased plant-disease attacks as well as diminished crop development (Conde and others 2008). Such climatological pressures caused losses for coffee producers of up to 36 percent in 1970, 1973, 1981, 1982, 1989, 1996, and 1997.

Michoacán

Michoacán is the state with the 12th largest indigenous population in Mexico: about 200,000 people out of 3.9 million inhabitants. There are three main ethno-linguistic groups in three clearly defined regions: the P'urhépechas in the center, the Mazahua-Otomí to the east, and the Nahuatl on the northern coast (figure 4.5).[7]

Severe climate change and variability are expected in the state of Michoacán during the next century (figure 4.6). Based on normalized temperature and precipitation data from 6,000 weather stations in Mexico, southern United States, Belize, and Cuba, the increased aridity projected for 2030, 2060, and 2090 will have dramatic effects on production and living conditions across the region in general and in central Mexico in particular.[8]

Mexican agriculture in general is vulnerable, and climate change and variability adds to this (Eakin 2005). Maize is highly vulnerable to changes,

Figure 4.5   Five Major Groups of Indigenous Peoples of Michoacán, Mexico

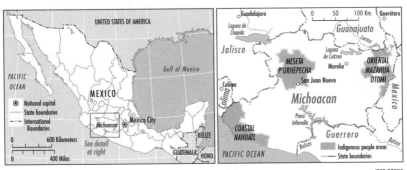

IBRD 37808
MAY 2010

Source: Elaborated from Comisión de Desarrollo Indígena (CDI).

and all projections show that areas unsuitable for maize production will expand by 18 percent if the mean temperature increases by 2°C and precipitation decreases by 20 percent. Throughout much of Mesoamerica, farmers grow many different varieties of maize, some of them associated with specific ethnic groups (Peralez, Benz, and Brush 2005). These varieties have specific climatic requirements. Traditional systems are dynamic—farmers experiment with different maize varieties. But the scale and pace of climatic change is unprecedented and may overwhelm the capacity of traditional maize knowledge/varieties to adapt. If the climate changes quickly, farmers may not be able to obtain traditional seed varieties that match their new growing conditions. They may shift to "improved" or high-yielding varieties, making them more dependent on the market and hastening the loss of agro-biodiversity, in addition to the loss of culturally preferred crop varieties and transmission of local knowledge. For a temperature increase of 2–3°C, negative effects are expected for more than 400,000 km2 of land where maize is currently grown.

The main adaptation strategies mentioned by farmers are (a) increased application of urea, (b) switch to cultivars with higher yields,[9] (c) irrigation, and (d) nonfarm employment and migration (Eakin 2003, 2005). The first three strategies depend on subsidies, access to water, and funds for investment, and they are considered too expensive for most farmers. This helps to explain why so many families in the Caribbean and Central America enhance or include nonfarm sources of income as an element in their coping and adaptation strategies. Many, for instance, resort to temporary and permanent migration for wage work and bring or send home funds.

Figure 4.6   Increasing Aridity in the State of Michoacán, Mexico, 2030, 2060, and 2090

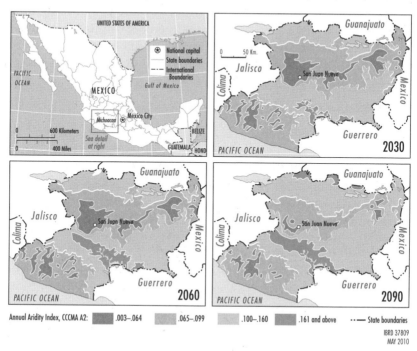

Annual Aridity Index, CCCMA A2:   .003–.064   .065–.099   .100–.160   .161 and above   ·—· State boundaries

IBRD 37809
MAY 2010

Source: Elaborated from Saenz-Romero and others (2009).

San Juan Nuevo Parangaricútiro, an indigenous community that we visited in September 2008, serves as a case in point for the general drying trend described above. The village is just east of the extinct Volcano Tancítaro (indicated by a circle on figure 4.6). The area produces rain-fed maize and avocados. Projected aridity is likely to severely affect these crops, to the point where the indigenous and other populations will have to switch from their current cultivars to more drought-resistant varieties (these have shorter growth periods and lower water requirements) or convert from maize to a less water-intensive crop, such as cassava, which is much lower in protein. The social result of this would be more poverty, because crops that grow well in an arid zone do not yield very much per hectare. The state government of Michoacán envisages its state as a source of water for the drought-ridden central states, although Michoacán, too, can expect aridity.[10] The community of San Juan Nuevo Parangaricútiro will probably have to find alternatives to its currently rain-fed orchards and maize production. Several community members express concern that the principal crop, maize, is already hard hit by unforeseen hailstorms, torrential rain, and prolonged periods of dry weather. The community has in recent years successfully reversed deforestation and reforested large areas through a participatory, though costly, process. If this effort can be kept up, and the forests adjust to the changing conditions, the community may overcome the threats of climate change and variability.

Indirect Social Impacts from Climate Change

Climate change and variability may have several indirect social impacts. As noted above, the indigenous peoples of the Nicaraguan Caribbean region believe that the main indirect impact from hurricanes and storms is a more rapid advance of the agricultural frontier. This advance has created ethnic frictions between indigenous groups and poor mestizo farmers because the latter are invading historical indigenous territories that the mestizo see as "uninhabited free land." In reaction to this potential ethnic conflict, indigenous communities have mobilized themselves to demand, from the government, the demarcation and titling of their traditional territories. As a result of this pressure, an indigenous territory demarcation law has been passed, but the process of actually titling the indigenous land has been rather slow.[11]

Another indirect impact is increased dependency on the market economy, as indigenous people now import more goods and materials that are not produced in the region. The sources of income to pay for these commodities are the commercialization of valuable natural resources on the international market (mainly shrimp and lobster) and the increased job opportunities for youth through migration and on cruise ships, with the consequent exposure to new life styles.

Summary

The physical consequences of changes in climate patterns are diverse, ranging from slow-onset phenomena, such as rising sea levels and melting glaciers, to increased intensity of extreme events, such as tropical cyclones and floods that occur suddenly and at variable intervals. These consequences affect people's livelihoods in different ways. Hurricanes, cyclones, and strong storms draw most of the attention regarding climate change and variability in the Caribbean and Gulf of Mexico, but the slow-onset processes associated with projected climate change will have consequences in this subregion that are at least as severe. Increased vulnerability can be expected for large populations, not least for the indigenous populations, with very similar impacts to those suffered in the Andes, sub-Andes, and the Amazon regions.

Large proportions of the most vulnerable people lack adequate coping strategies of their own. Agriculture is by far the most important source of income for indigenous peoples in the region; hence, the susceptibility of agriculture to climate change and variability directly affects indigenous peoples' livelihoods. It is therefore vital that adaptation strategies addressing this vulnerability are effective. Slow-onset hazards do not always demand humanitarian intervention, particularly when governments and communities work together to reduce the impact on affected people. Even when intervention is necessary, it is important to remember that many communities have been living with periodic or cyclical drought for generations. Consequently, they have developed ways to cope—strategies that differ among groups (pastoralists, agropastoralists, farmers) and within groups, from richer to poorer. Particularly during the early stages of drought, humanitarian efforts should aim to support these coping strategies, thereby strengthening a community's resilience. Continued and sometimes strengthened political and other support for poor producers is needed: for example, investments; subsidies; access to water and land; and responsive, reliable, and relevant knowledge and information. Another example could be support for farmer-to-farmer seed networks for traditional maize varieties.

By definition, there is more time to plan and implement an appropriate response to a slow-onset hazard such as drought. Yet evaluations still criticize the apparent lack of learning and the repetition of mistakes, including the fact that the humanitarian system often does not intervene until the crisis stage.

Notes

1. While sea-level rise has received considerable public attention, this aspect of climate change is not covered in this chapter because the sea-level rise for this region is not projected to be significant within this century.

2. Alexis Jones' thesis, "Rural Households' Vulnerability to Intense Rainfall Events in Chiapas, Mexico," investigated whether people who received early warnings on the advent of Hurricane Stan took steps to reduce hurricane impacts. He found that even though they had prior experience with hurricanes, they did not take these steps-the forecasts were confusing (even for people who understood Spanish), the forecasts changed, and the people could not mobilize a great deal of resources on short notice to enable them to respond (Jones 2000).

3. http://www.adapcc.org.

4. Migration patterns also reflect political violence in both Chiapas and Guatemala.

5. AdapCC program in Mexico, Nicaragua, Peru, and Tanzania.

6. Coffee growing requires maximum summer temperatures of 26°C and in excess of 2,000 millimeters of annual precipitation.

7. The number of indigenous inhabitants is defined by language. This method usually provides an underestimate (Hall and Patrinos 2006).

8. Projections by the Intergovernmental Panel on Climate Change using the *Special Report on Emission Scenarios* A2 criteria (slow and differentiated technological development).

9. The issue of cultivar is a much-investigated field of research (including by the International Maize and Wheat Improvement Centre, CIMMYT) for maize. Higher yields are often linked to market dependence, and traditional varieties are more drought-resistant and often more resistant to "lodging" (being blown over by wind).

10. Personal commentary, Secretary for the Environment, Michoacán (2008).

11. In Nicaragua, this pressure escalated to a lawsuit by an indigenous community against the Nicaraguan government at the Inter-American Court.

References

Alfaro, Julio, Sergio Saladaña, and Kerstin Linne. 2008. "Sistematización del Análisis de Riesgos y Oportunidades (ARO)." Mascafé, México. *AdapCC.* July–September. http://www.adapcc.org/download/Sistematizacion_ARO_Mexico_20080923.pdf.

Balée, William. 1989. "The Culture of Amazonian Forests." In *Advances in Economic Botany*, vol. 1 of *Resource Management in Amazonia: Indigenous and Folk Strategies*, eds. D. A. Posey and W. Balée, 1–21. Bronx, NY: New York Botanical Garden.

Bebbington, Anthony. 1999. "Capitals and Capabilities: A Framework for Analyzing Peasant Viability, Rural Livelihoods, and Poverty." *World Development* 27(12): 2021–2044.

Bourdieu, Pierre. 1986. "The Forms of Capital." In *Handbook of Theory and Research for the Sociology of Education*, ed. J. G. Richardson, 241–258. New York: Greenwood Press.

Bradford, David. 2002. "Ecologia y medio ambiente en la Costa Caribe de Nicaragua: Descripción y manejo de ecosistemas tropicales." CIDCA, Universidad Centroamericana. Managua, Nicaragua: Multigrafic.

Carpenter, G. 2006. "Tropical Cyclone Review 2005." Instrat Briefing, January. http://gcportal.guycarp.com/portal/extranet/popup/pdf/GCBriefings/ Tropical_Cyclone_Review_2005.pdf.

CDI (Comisión de Desarrollo Indígena). 2008. Mexico City. Mexico.

CEPAL (Comisión Económica para América Latina y el Caribe). 2006. *Panorama Social de América Latina 2006*. Publicaciones de las Naciones Unidas, LC/G.2326-P/E. Santiago, Chile: United Nations. http://www.eclac.org.

Conde Àlvarez, Cecilia, and Sergio Omar Saldaña Zorrilla. 2008. "Análisis de Riesgos Climáticos para Productores de Café en México." Draft. Centro de Ciencias de la Atmósfera de la Universidad Nacional Autónoma de México (UNAM).

Conde, C., M. Vinocur, C. Gay, R. Seiler, and F. Estrada. 2008. "Climatic Threat Spaces in Mexico and Argentina." In *Climate Change and Vulnerability*, eds. Neil Leary, Cecilia Conde, Jyoti Kulkarni, Anthony Nyong, and Juan Pulhin, 279-306. Trieste, Italy: Academy of Sciences for the Developing World.

DFID (U.K. Department for International Development). 2001. "Sustainable Livelihoods Guidance Sheet." http://www.nssd.net/pdf/sectiont.pdf.

Eakin, H. 2003. "The Social Vulnerability of Irrigated Vegetable Farming Households in Central Puebla." *Journal of Environment and Development* 12 (4): 414–429.

Eakin, H. 2005. "Institutional Change, Climate Risk, and Rural Vulnerability: Cases from Central Mexico." *World Development* 33 (11): 1923–1938.

Hall, Gillette, and Harry Anthony Patrinos, eds. 2006. *Indigenous Peoples, Poverty, and Human Development in Latin America*. New York: Palgrave.

IWGIA (International Work Group for Indigenous Affairs). 2008. *The Indigenous World Yearbook*. Copenhagen: IWGIA.

Jones, Alexis, 2007. "Rural Households' Vulnerability to Intense Rainfall Events in Chiapas, Mexico." Master's thesis, University of California, Davis.

Kronik, Jakob. 2010. *Living Knowledge—The Making of Knowledge about Biodiversity among Indigenous Peoples in the Colombian Amazon*. Saarbrücken, Germany: Lambert Academic Publishing.

Layton, Heather Marie, and H. A. Patrinos. 2006. "Estimating the Number of Indigenous People in Latin America." In *Indigenous Peoples, Poverty, and Human Development in Latin America*, eds. Gillette Hall and H. A. Patrinos. New York: Palgrave.

Nitlapan, Universidad Centroamericana (UCA). 2007. Mairena Cunningham, Eileen. Gestión De Recursos Naturales en 16 Comunidades Indígenas de La Costa Caribe de Nicaragua. Eileen Mairena Cunningham con la colaboración de Renee Polka y Claribel Ellyn Gómez. (NIC S930.N5M29). Managua: Nitlapán.

Peralez, Hugo R., Brice F. Benz, and Stephen B. Brush. 2005. "Maize Diversity and Ethnolingusitic Diversity in the Chiapas, Mexico." *Environmental Sciences* 102 (3): 949–954.

Sáenz-Romero, Cuauhtémoc, Gerald E. Rehfeldt, Nicholas L. Crookston, Pierre Duval, Rémi St-Amant, Jean Beaulieu, and Bryce A. Richardson. 2009. "Spline Models of Contemporary, 2030, 2060 and 2090 Climates for Mexico and Their Use in Understanding Climate-Change Impacts on the Vegetation." Climatic Change, Springer, published online November 12.

SAGARPA (Government of Mexico: Secretaría de Agricultura, ganaderia, desarrollo rural, pesca y alimentación). 2008. http://www.sagarpa.gob.mx/dlg/chiapas/agricultura/Perennes/cafe.htm.

SEGOB (Secretaría de Gobernación de Mexico). 2002. See Conde Àlvarez and Zorrilla 2008.

UNDP (United Nations Development Programme), Nicaragua. 2005. *Informe de Desarrollo Humano 2005. Las regiones autónomas de la Costa Caribe ¿Nicaragua asume su diversidad?* 1st ed. Managua, Nicaragua: Programa de las Naciones Unidas para el Desarrollo.

U.S. Central Intelligence Agency. 2009. *World Fact Book.* https://www.cia.gov/redirects/factbookredirect.html.

U.S. Department of Commerce. 2009. National Climate Data Center. http://www.ncdc.noaa.gov/oa/climate/climatedata.html.

CHAPTER 5

Indigenous Peoples and Climate Change Across the Region

Table 5.1　Elements of a Comparative Framework on the Social Impact of Climate Change on Indigenous Peoples of LAC

Ecogeographical regions	Threats	Direct impact	IP adaptation strategy	Recommendations
Colombian Amazon: Amazonian dieback, less predictability and intensity of drought and precipitation patterns.	• Seasonality varies. • Temperature increases. • Prolonged droughts. • Unpredictable precipitation. • Greater autonomy, more aware, more vulnerable. • Less autonomy, less aware, more assimilated.	• Changing seasonality affects production and social institutions, relations, and livelihoods around it. • Food availability, due to (a) diminishing availability of fish, (b) less game, and (c) less agricultural produce.	• Rely on pluriactivity in diverse ecological environments. • Experiment with modified horticultural practices. • Slash and burn when a few dry days occur; manually built fires. • Open gardens in secondary forest (shorter dry periods).	• Contain degradation of natural resources, including deforestation, while upholding traditional indigenous people (IP) institutions. • Promote the active participation of IP communities in developing adaptation initiatives. • Protect IP rights to land and natural resources through demarcation and titling of IP territories.
The Caribbean – Mexico Nicaragua: Increased intensity of extreme events like storms and hurricanes and slow-onset climate impacts.	• Increased intensity and frequency of extreme events. • Intense precipitation followed by droughts. • Vulnerability linked to access to land, resources, types and quality of livelihood strategies, and institutional affiliations.	• Poor communities lost access to resources (fishery, forestry, and farming), infrastructure, and personal belongings. • Urban poor (mainly IP and mestizos) affected by destroyed infrastructure and belongings. Few are insured.	• Combine different productive activities: agriculture, forest products extraction, and sand mining combined with fisheries and seasonal work in cruise ships. • Adaptation depends on location, disposition, and quality of infrastructure. • Higher social cohesion of IP communities is positive.	• Design early-warning systems on hurricane development with reach to IP and poor farmers. • Implement awareness campaigns and disaster risk-reduction training. • Support government emergency and disaster risk-reduction plans; construction codes for disaster-prone areas.

Region	Observed changes	Effects	Adaptation strategies	Recommendations
 **Bolivian/Peruvian Altiplano Andean Region:** Increased temperature, with glacier retreat and highland upstream changing of water balances.	• Increasing temperatures, glacier retreat, rains, hailstorms, frost, droughts, and changes in seasonality. • Absence of water. • The poorest in rural areas (mostly IP), who depend on affected NR, changes in agricultural calendar.	• Decrease in crop profitability. • Losses in production. • Plagues on potato and onion crops. • Changes in making of *chuño*. • Loss of wildlife. • Diseases in cattle. • Skin sunburns.	• Search for colder places in high-elevation areas to produce *chuño*. • Search for water in more distant places. • Have a sustainable community development plan to look for new productive alternatives.	• Strengthen institutions, including municipalities, through community participation and radio broadcast info on adaptation projects. • Increase local capacity to "read and interpret" local climate change; improve local access to climate-change information and analyses.
 **Bolivian/Peruvian Sub-Andean Region:** Increased temperature, glacier retreat, and warmer mid-downstream changing water balances.	• Increased temperature, glacier retreat, seasonal variation. • More intense rains, hailstorms, and floods. • Poor IP and mestizos from change in the agricultural calendar, unpredictability, erosion, soil salinization, and so forth.	• Less water availability. • Less profitability of crops. • Losses in production. • Plagues in citrus fruits. • Agricultural calendar changed. • Mosquitoes found in new area. • Sunburn. • Health at increased risk.	• Elevate river banks to avoid flooding. • Increase coca production. • Change of income source to tourism. • External and internal temporary migration. • Search for training resources, projects, and institutions.	• Strengthen local institutions to increase adaptive capacity, including development of local agriculture (seeds, methods, and so forth). • Design rural development plans inclusive of IP participation. • Study potential impact of climate change on IP rural-to-urban migration.

Source: Elaborated based on field interviews. Finding presented in ch. 2–4.

Across the region, indigenous peoples already perceive and suffer effects from global climate change and variability, some of which cause serious problems. This chapter draws on the three regional case studies to provide a comparative analysis of the impact of climate change on rural indigenous communities in LAC. Based on fieldwork in Bolivia, Colombia, Mexico, Nicaragua, and Peru, the chapter examines the threats from climate change and variability, the impacts of these threats on indigenous communities, and the adaptation strategies applied by various indigenous groups (table 5.1). The chapter concludes with a discussion on the possible contributions of indigenous peoples to climate-change mitigation strategies.

Indigenous peoples across the region embody great diversity. While the findings for each of the case studies presented here are specific to the communities researched, different indigenous groups within the same eco-zone may experience common threats and impacts from climate change. Adaptation strategies, however, are likely to vary between different peoples relative to the traditional knowledge and cultural practices of each community. Therefore, it is important to keep the sensitivities of the specific groups in mind when working on climate-change adaptation in indigenous communities. Operational recommendations are offered in chapter 6.

Threats

As has often been stated, communities that depend directly upon the use of natural resources for their livelihoods are vulnerable to climate change (Diaz 2008; Gerritson 2008; Smith 2008; Sulyandziga 2008). Like other people who depend on natural resources, rural indigenous communities face threats from unpredictable seasonal variations, rising mean temperatures, droughts, increased amounts and intensity of precipitation (rains, hailstorms, and floods), frosts, storms, and hurricanes.

But many indigenous communities are disproportionately vulnerable to such threats. For members of these communities, survival often depends to a very high degree on traditional knowledge systems, evolved over generations, and on direct observation and interpretation of the natural world—which they perceive to be increasingly in disarray as the result of changes in seasonality and increases in extreme weather. Meanwhile, the economic resources they depend on for their livelihoods are already under threat from societal trends such as deforestation, advancing colonization, and expansion of towns and commercial agriculture, with associated demands for water. Further, other important changes taking place in their ways of life—as the result of increased contact with the market economy, school education, mass media, and a range of external agents—have tended to erode their traditional knowledge about how best to manage in the face of changes in natural phenomena. Indigenous people's ability to cope with these complex trends differs from place to place; it is framed by their own cultural traditions and by relations with state and nongovernmental agents, as well as increasingly by mainstream education and by the media.

As would be expected, many of the threats from climate change and variability vary among ecological regions and communities, but several are common across the continent. Probably the most widespread and serious of these is the disturbance of the annual ecological calendar and the associated disruption of agriculture, hunting, and gathering. A recurrent lament heard during all our field visits was that seasonal variations have become so unpredictable that indigenous peoples' adaptation strategies, developed to tackle the normal span of variation, no longer provide the necessary security. In all the case-study countries, changes in the timing of the dry and rainy seasons, alterations in the flood pulses of the rivers, winds, and abnormal cold and heat have become apparent during the last decade—and increasingly since 2005. The seasonal rhythm is crucially important; it orders the timing of the cropping and livestock cycle and the ritual practices that indigenous people use to prevent illnesses and promote well-being; it also is crucial for the reproduction of wildlife. Its disruption jeopardizes food supplies and undermines the array of solutions provided by the indigenous cultural institutions.

The major interregional differences we observed with respect to threats from climate change stem from ecogeographical conditions. In the Colombian Amazon, indigenous people perceive big changes in seasonality: ecological markers are occurring abnormally early or late, decoupled from the weather or season they used to mark; and they differ in timing, kind, or intensity from the normal recurring interannual variations. In lowland Amazonian forests, drought is exacerbating the effects of ongoing deforestation, increasing the risk of forest fires, and threatening a collapse of the rain forest.

In the Andean and sub-Andean regions, increasing mean temperatures are causing glacier retreat; variations in seasonality and in the amount and intensity of rains, hailstorms, and frosts during unusual periods of the year; and droughts. All of these changes affect the Aymara and Quechua peoples in these regions, putting food security at risk, affecting social stability, health, and psychological well-being. In the high Andes, livelihoods are threatened by the shrinking of the glaciers; with less snowfall stored as ice, people and their livestock herds who traditionally depended on water from snowmelt now suffer water scarcity in the prolonged dry seasons. Also particular to the high mountains are the risks to crops that are posed by unexpected hailstorms and frosts. For indigenous people in the lower Andes, the melting of glaciers presents different threats, including floods and soil erosion.[1]

The Mesoamerican and Caribbean region is now notorious for the increased intensity and frequency of natural disasters, and the region harbors some of the world's countries most vulnerable to the effects of climate change and variability. Projections show an increasing risk of more powerful hurricanes and a continuing upward trend for intense precipitation, followed by extended periods of drought stemming from higher temperatures. Such hazards threaten indigenous people just as they do other people. But certain characteristics of indigenous people's traditional social organization, culturally based knowledge and practices, livelihood strategies, and degree of access to assets may make for differences in their levels of preparedness, types of responses, and needs for support, as seen in cases in Nicaragua and Mexico. Meanwhile, slow-onset aspects of climate change are endangering entire ecosystems in this part of the LAC region: gradual warming and acidification of the oceans threatens the viability of coral reefs and mangroves, which some communities rely on for their livelihoods, and drought threatens the continued viability of traditional cropping in many areas. In coffee-growing areas of Chiapas, Mexico, for example, most scenarios project decreased precipitation. Smallholder coffee producers there, most of whom are indigenous people, have already lost adaptive capacity as the result of soil degradation, lack of water, and natural hazards (frosts, torrential rains on high-altitude hillsides, and drought at lower altitudes). The risk now is that coffee productivity will decrease dramatically, and production may even disappear in the lowlands, resulting in migration toward higher altitudes and into the jungle areas, putting pressures on the Lacandon Maya peoples.[2]

Impacts

Across the region, the changes in precipitation patterns and seasonal regimes disrupt the agricultural calendar and the availability of food foraged from the wild, with serious consequences not only for indigenous people's food security, but also for their health conditions and cultural identity. The changes affect crop production and the availability of fish, wild fruits, and game, and they increase the incidence of livestock diseases that impair the livelihoods of the many indigenous people who are pastoralists.

Human health is widely perceived as a key area of impact. Particular concerns are the spread of disease vectors into areas where they could not previously thrive (in some Andean regions), increased incidence of respiratory and diarrheal diseases (in Amazonian areas), and widespread increased difficulty in obtaining adequate nutrition. In turn, malnutrition has consequences for people's ability to resist infectious diseases, and it compromises the development of children.

The major interregional differences in the impacts observed are between areas prone to rapid-onset hazards and those prone only to slow-onset hazards. In Mesoamerica and the Caribbean, increasingly severe storms and hurricanes damage and destroy infrastructure and personal property and decrease access to the resources needed for fisheries, forestry, and agriculture. Indigenous peoples particularly affected by climate change and variability include the Miskitu, the Sumu-Mayangna complex (Ulwas, Twaskas, and Panamaskas), and the Rama, all of whom belong to the Macro-Chibcha linguistic family; and other ethnic communities such as Creoles and Garifuna (both Afro-descendants). A finding that needs further systematic inquiry is indigenous people's apparent loss of land as the indirect result of hurricanes: in Nicaragua, for example, hurricane damage to previously impenetrable forest has made it easy for settlers and cattle ranchers to encroach into indigenous territories.

In indigenous communities studied in the Colombian Amazon, changes in precipitation and seasonality have direct immediate effects on human activities and health. Often crops now fail repeatedly, limiting the diversity of sources of nutrition when only the toughest types of crop survive.[3] River fish and turtles are important sources of food for indigenous communities, but their reproduction has been badly damaged because the rivers no longer rise and fall seasonally as they used to. The greatest current concern of indigenous peoples in the Amazon region, however, is their social situation. The traditional harmony of their lives with nature is disturbed not only by climate change and variability, but also by the effects of advancing colonization, destruction of the forest, political unrest, illegal coca cultivation, excessive resource exploitation, gold mining, and trade. In combination, these changes lead to increasing social disarray.

In the Andean regions, field research shows a long list of negative impacts: worsening scarcity of water for crops and livestock; erosion of ecosystem and natural resources, for example, through salinization of soils; changes in biodiversity as a consequence of the spread of alien species; unfamiliar and more virulent plant diseases and pests affecting crops; crop losses; a higher death toll among livestock; and higher risks of infectious diseases. These effects lead to human and material losses and threaten food security, both within indigenous villages and among the populations who depend on the food produced by these villages. Likewise, the negative impacts induce migration to cities, where people crowd into poor, conflict-ridden neighborhoods on urban fringes. Changes taking place in the geographical range of crops as the result of warming temperatures are not all detrimental, however. In Peru, for example, it is widely reported that maize and, in a few areas potatoes, can now be grown at higher altitudes, improving some people's livelihoods.

Just as for nonindigenous people, some groups of indigenous people are more vulnerable than others, reflecting differences in their social and ecological situations, including access to land and other natural resources, type of livelihood strategy, degree of contact with mainstream society and integration with the market, cultural resourcefulness, and institutions. In some respects, age and gender are important factors. For many indigenous people, the impacts from climate change are particularly damaging because they are compounded by other factors that have been increasingly eroding the traditional ways of life.

Our interviews showed that the indigenous peoples most aware of and most vulnerable to climate change and its effects are those within the Amazon basin who have greater territorial autonomy. In this category are the indigenous groups of the Vaupes area (Eastern Tukano, Maku-Puinave, and Arawak linguistic stocks), the Caqueta-Putumayo region (Witodo, Bora-Miraña, and Andoke stocks), the Guainia region (Maku-Puinave and Arawak stocks), and some groups of the Amazon Trapeze (Tikuna, Peba-Yagua, and Tupi stocks). Though these peoples have contact with mainstream society, are incorporated in some measure into the market economy, and have access to public health and education services, they derive their livelihood mostly from forest and water resources. They depend heavily on gardens slashed in mature forest and planted with a large variety of species. For protein, they rely largely on fish and game, and they maintain health through their own means and knowledge. They maintain an active and engaged ritual life. Their livelihood rests on their ability to interpret regular natural cycles and act accordingly, but their knowledge and practices increasingly fail to respond effectively to the changes in precipitation patterns.

Indigenous peoples of the Amazon who have restricted territorial autonomy and little or no access to mature forest are less vulnerable than the more traditional groups, because they have greater access to sources of income and information that do not depend on the direct observation and interpretation of nature. This broad category of indigenous peoples includes those living at the fringes of the colonization areas in the foothills of the Andes and along some of the main rivers of the Amazon basin, such as the Guaviare River, and those living close to urban areas. They are impacted by climate change to the extent that they use river and forest resources, but they are less in tune with the seasonal calendar (or their awareness is restricted to those aspects that directly affect their activities), their traditional knowledge is more limited, and ritual specialists generally play a smaller role in their lives. They depend on horticulture in secondary forest, cash crops on alluvial soils, commercial fishing, wage labor, tourism, and sale of handicrafts for their livelihood. The combined effects of climate change and variability tend to prompt or accelerate their reliance on wage jobs, market integration, and migration to urban areas—and to cause their impoverishment.

High-elevation Andean herders are less isolated from mainstream society than some Amazonian groups, but they constitute another group that is severely affected. Herders use the Andean wetlands as permanent humid pasturelands, or *bofedales*, to feed their livestock (llamas, alpacas, and vicuñas) during the dry season. Some watersheds have already disappeared, according to herders, who are already moving their cabins to higher altitudes for herding, looking for water sources (Gil 2008). Rhoades, Rios, and Ochoa (2008) document how current water concessions based on outdated flow figures create conflicts over increasingly scarce water.

Women within indigenous communities across the region carry a heavy burden of impact, as they are often traditionally in charge of the routine labor in horticulture, while men are often the ones to clear the forest itself, involving heavy labor but much less of a time demand. Work in open fields under rising temperatures, crop failure and replanting, and lower crop yields affect women's health and psychological well-being. Making gardens in secondary forest, or half-burnt gardens in mature forest, requires more labor than traditional gardens in mature forest to eradicate weeds and to manually complete the burning process. Children's ill health and malnutrition directly affect women's work and responsibilities.

Elders and traditional leaders—who are the experts within traditional knowledge systems—lose credibility and have their authority impaired when weather becomes impossible to predict. Frost, harsh rain, or drought have been arriving at times so unpredictable that main staple crops have been ruined. When nature's cycles become erratic, traditional authorities cannot guarantee abundance and prosperity. The unpredictable changes in the agricultural calendar also make it increasingly difficult for elders, ritual specialists, and healers to grow the complex range of plants they need for medicinal purposes and ceremonial activities. Meanwhile, new and stronger diseases appearing in their communities outpace their abilities for prevention and healing. These impacts affect their psychological well-being and their authority, with consequences for their social organizing capacity relevant to production, reproduction, and leadership.

Thus, when unpredictable natural events happen repeatedly, they undermine ritual practices and the joint social memory, and people lose confidence in the ability of their elders and traditional leaders to restore the necessary balance between the human, natural, and cosmological realms. The status of the leaders fall and people look elsewhere for solutions to their problems, both by turning to other bodies of knowledge and by migrating away in search of a more secure livelihood. Cultural practices that have evolved in synchrony with nature's cycles lose their significance and symbolic power. This leads to the reduction of a people's cultural capital, diminishing cohesion and the resistance to divergent belief systems, and may in turn lead to a reduction of social interactions and hence a reduction of social capital. Conflicts arise between traditional authorities and young political leaders who seek new ways to sustain their people.

People interviewed across the LAC region referred to these changes as leading to the end of a way of life that has sustained their communities for generations. The more traditional among them interpreted the alteration of natural cycles, and the failures in their livelihood and health, in a moral and ethical framework with a sense of shared responsibility. Interviewees had often heard, both in the news and through visitors, about global climate change and variability. In places as different as the Andean Altiplano, the Amazon, the Mesoamerican highlands, and the Caribbean coast, they clearly stated that while "white people" may have caused damage to the planet, indigenous peoples share the responsibility for their failure to live and manage life as ordained by their Creator.

Potential Effects

The potential effects of climate change are uncertain, given the many interrelated variables within social, natural, and economic systems—and unforeseen events. There is insufficient understanding of how chain reactions may reduce the resilience of some systems to change, while other systems more easily adapt. Social and biophysical systems can reach so-called "tipping points," beyond which there are abrupt, accelerating, or potentially irreversible changes. The United Nations Environment Programme's *Global Environment Outlook* Report (UNEP 2007) discusses scenarios of the increasing risk of crossing these tipping points along with processes of environmental degradation. Nepstad and others (2008) review the dynamics of a situation where fire that is used to prepare land for agriculture spreads into surrounding forest areas, causing their degradation; this can potentially push a whole region across such an ecoclimatic tipping point, degrading it irreversibly from forest to scrubland. By first considering the projected climatic changes, we can try to extrapolate some of the related social impacts.

Temperatures. As discussed in chapter 1, climate-change projections imply that over most of the LAC region, temperatures in all seasons will continue to rise during the 21st century. Heat waves are likely to be more frequent and intense, and the higher temperature level in general will tend to favor a longer warm season with possible related extreme events such as hurricanes.

The high Andes are projected to experience a greater temperature rise than the average for the LAC region. Glaciers are their single major seasonal store of water. The effects of rising temperature on the glaciers, and the activities that depend upon this water reserve during the prolonged dry seasons, can be expected to be severe for farmers and herders, as well as for urban water consumers. Processes of social and cultural disintegration already seen in the Andes are likely to increase further if effective mitigative and adaptive measures are not implemented.

In the Amazon region, the expected higher temperatures will worsen the destructive effects of deforestation and increase the risk of wildfires (Nepstad and others 2008), which in turn will add pressure on indigenous peoples' lands and resources and may lead to increased colonization by settlers.

Precipitation. Broadly, as described in chapter 1, dry areas are projected to become dryer and wet areas wetter. It is uncertain how annual and seasonal mean rainfall will change over northern South America, including the Amazon forest. Changes in atmospheric circulation may induce large local variability in precipitation in mountainous areas, making the right adaptation strategies even more difficult to determine.

Extreme events. Available models clearly suggest that changes will take place over this century that will generally be in the direction of more of the extremes, that is, more intensive precipitation, longer dry spells and warm spells, heat waves with higher temperatures than generally experienced up to now, and a larger number of severe hurricanes. Further increases are expected in floods and droughts and in the intensity of tropical cyclones. In the Caribbean and the American isthmus, hurricanes affect indigenous communities as hard as other poor rural groups. In some cases, social cohesion has helped indigenous communities to adapt, but in other cases indigenous communities have lost everything and have needed to relocate or to adopt new livelihood strategies. More intense extreme events are likely to exacerbate these effects.

Adaptation

Across the continent, indigenous peoples are making a range of adjustments in their productive activities to cope with the effects of climate change and other new influences. Though poor people who depend heavily on natural resources for their livelihoods typically have little capacity to adapt when their livelihoods are threatened (World Bank 2000), their resilience may be enhanced by well-functioning sets of regularized practices (Robledo, Fischler, and Patino 2004; Salinger, Sivakumar, and Motha 2005; Thomas and others 2007). Over generations, indigenous communities have developed and actively maintained such practices, which are a key common identifier of their well-being and their capacity to adapt to social and other changes.

In many cases, however, we found that the kind, frequency, intensity, and unpredictability of the effects of climate change and variability leave indigenous communities with a sense of frustration, and in some cases even resignation and acceptance, with little knowledge about how to act or who could help. Some of the communities visited are being compelled to change their livelihoods so dramatically that they lose vital conditions for the development and reproduction of their culture. Increasingly, their members are migrating away in search of education and seasonal wage jobs, and some are uprooting permanently, as discussed further below. Their traditional knowledge, institutions, and practices may be rendered superfluous, temporarily forgotten, or completely lost. For example, in the Andean region, we found that some community members have knowledge that will allow them to foresee and be prepared for some climate events, but the cultural institutions that normally distributed and validated this knowledge are now weak or not being used sufficiently, so the knowledge benefits only a few people or is in danger of disappearing.

The fact that indigenous peoples often live in an institutional vacuum, without any linking social capital to institutions outside their own society, underlines the importance of taking the political and governance dimensions of their adaptation seriously. Thus, as argued at the end of this chapter, it is vital that further strategies be developed to help indigenous communities to adapt to climate change and variability.

Adaptation strategies vary according to the opportunities at hand. In the coastal Caribbean areas studied, people are very attached to their communities, but the dwindling of natural resources, exacerbated by damage from climate change and variability, and the lack of job opportunities make young people leave for temporary work elsewhere. In areas of Mesoamerica where farmers grow many different varieties of maize, traditional cultivation systems are dynamic, but the scale and pace of temperature change are unprecedented and may overwhelm the capacity of traditional maize knowledge and varieties to adapt. For many farmers in this situation, the main alternative options are out of reach: increased use of urea, a switch to higher-yielding cultivars, or use of irrigation depends on subsidies, access to water, and funds for investment. Thus many are resorting to temporary and permanent migration for wage work, combined with bringing or sending home funds.

In Andean and sub-Andean regions, indigenous people's adaptive strategies include the introduction of new crops, the revival of old customs and rituals, and temporary migration. Aymara communities in the Altiplano are shifting into new crops such as irrigated onions, changing the varieties of potatoes they grow, adopting the use of pesticides, and using cropping techniques learned through training courses. In the sub-Andean region, Aymara and Afro-Bolivian communities are traveling further to obtain the water they need, increasing their production of coca to compensate for losses in productivity of other crops, and shifting from market agriculture to new income-generating activities such as tourism. People from both these regions are migrating to other places to obtain wage work, although some return to their lands the following year to produce again. Interviewees in the sub-Andean region expressed a greater interest in tourism and in receiving support from NGOs and government programs to address climate change and variability.

In both the Andean regions, indigenous people will need to further adjust their livelihood strategies to the changing of water flows. Some have no cultural adaptation strategies for this purpose, and the limited presence of state institutions adds pressure on the capacity of local authorities. The pressure from the decreased availability of water that results from climate change—combined with an institutional vacuum, loss of esteem for traditional authorities, and a loss of trust in traditional knowledge—threatens the viability of traditional culture and religion. The difficulties of traditional indigenous institutions in meeting the needs of their community members lead to increased migration and social change.

Across LAC, migration of indigenous people is increasingly common—including permanent migration of both individuals and entire households. Not only do individual family members leave, but whole families are now uprooting. Some communities are left ghostlike; in others, only children and the elderly remain. Moving closer to urban areas may provide greater access to the market and to public health and education services, which can help to buffer the impacts of climate change on the part of their livelihood derived from natural resources, but often it leads only to further impoverishment.

In some cases, indigenous communities have shown spontaneous and intuitive capacity to adapt to the effects of climate change. Previous experiences with training projects, as well as technical assistance in agriculture and other areas, have indirectly helped them to identify alternatives and hence increase their capacity to adapt.

Also helpful have been indigenous cultural institutions as regularized practices that shape the management of natural resources (North 1990; World Bank 2002, 2006b). As seen from examples in Colombia and Mexico, indigenous peoples often use a broadly based, multiple-use, adaptive approach for managing available resources, "learning to do everything to survive." Using data obtained from 63 household interviews in three Yucatec Maya communities located in the northeastern portion of the Yucatan Peninsula, García-Frapolli, Toledo, and Martínez-Alier (2008) show that contemporary Yucatec Maya households manage up to five land-use units[4] in which they implement 15 different economic activities. Some of their livelihood sources are under severe pressure from climate change and variability; success in replacing them (for example, with ecotourism) was found to depend upon the degree to which the new activities could be supported by and integrated into their traditional approaches to managing natural resources—as well as on a consistent approach to governance and on political will.

Thus far, only limited knowledge is available on the role of cultural institutions in the context of climate change and variability. Cultural institutions of indigenous peoples can play an important role in the context of adaptation—promoting the incorporation of experience with previously unknown phenomena (changing climate and, thus, flora and fauna)—and perhaps also in climate-change mitigation through forest protection policies, as discussed in the following section.

Indigenous Peoples, Mitigation of Climate Change, and Protection of Biodiversity

Reducing deforestation and forest degradation is the forest-related climate-change mitigation option with the largest global short-term impact on the carbon stock per hectare and per year, because large carbon stocks are not emitted when deforestation is prevented (IPCC 2007). Curbing deforestation is considered an immediate and highly cost-effective way of reducing greenhouse gas emissions at a significant scale because it does not imply the development of new technology, except perhaps technology for monitoring (Stern 2007). Moreover, it is assumed, forest-related mitigation options can be designed and implemented to be compatible with adaptation, and can produce substantial co-benefits in terms of employment, income generation, biodiversity, watershed conservation, renewable energy supply, and poverty alleviation (Stern 2007).

As Latin American countries develop and negotiate efforts to mitigate climate change, at the local as well as the global level, some instruments originally developed for other objectives—such as dams, biofuels, or forest protection schemes—are being rediscovered within the climate-change debate and are being characterized both as solutions and as reasons for concern. Growing international recognition of the stewardship role of indigenous peoples has been hard-won by indigenous peoples' organizations and advocates, who claim indigenous peoples' rights to territory, land, and resources based on the beneficial effect of these rights on biodiversity. Here we examine the important contributions that indigenous peoples can make to climate-change mitigation by acting as stewards of natural resources and biodiversity in the territories they live in, provided that their rights are recognized and respected.

Areas governed by indigenous peoples are less prone to deforestation than others. Reserves are highly effective in protecting forests. Through their reserves, indigenous peoples are as effective as the state parks in protecting land from deforestation and fires in less-contested areas (Nepstad and others 2006) and better at policing the forests in highly contested areas Adeney, Christensen, and Pimm 2009). For example, in Brazil, areas governed by indigenous peoples strongly inhibit deforestation and forest fires along the agricultural frontier.[5] The inhibitory effect of indigenous people's stewardship on deforestation is strong even after centuries of contact with mainstream society. Indigenous lands occupy one-fifth of the Brazilian Amazon—five times the area that is under government protection in parks—and are currently the most important barrier to Amazon deforestation. In nonforest ecosystems, carbon is sequestered in high-elevation grasslands, which indigenous peoples manage effectively, and in marsh wetlands.

Supporting indigenous peoples' rights may help governments to achieve targets for reducing carbon emissions. In LAC, the burning of forests is one of the main sources of greenhouse gas emissions. As part of their efforts to mitigate the rate of climate change, several LAC countries are launching programs to reduce emissions from deforestation and degradation (REDD).[6] In the LAC context, it is difficult to imagine much REDD without indigenous peoples' participation, simply because they control and often own large tracts of dense forest. If forest protection is to be an effective climate-change mitigation mechanism in LAC, indigenous peoples' rights must be recognized when designing and negotiating agreements for this purpose. Indigenous peoples fear having their autonomy and authority undermined by entering into government-negotiated REDD agreements.

In April 2009, the Indigenous Peoples' Global Summit on Climate Change adopted the Anchorage Declaration, which calls upon "the UNFCCC's decision-making bodies to establish formal structures and mechanisms for and with the full and effective participation of indigenous peoples."[7] The Anchorage Declaration further states, "All initiatives under Reducing Emissions from Deforestation and Degradation (REDD) must secure the recognition and implementation of the rights of indigenous peoples, including security of land tenure, recognition of land title according to traditional ways, uses, and customary laws and the multiple benefits of forests for climate, ecosystems, and peoples before taking any action."

The negotiated text of the June 2009 preparations in Bonn for the Conference of the Parties to UNFCCC in Copenhagen (COP15) outlines a possible scope for policy approaches and positive incentives on issues relating to REDD and the role of conservation, sustainable management of forests, and enhancement of forest carbon stocks in developing countries.[8] The Bonn text stresses that actions should be country driven and subject to national context. However, the Bonn text places emphasis on broad participation, so that indigenous peoples and local communities are involved in the implementation of the actions and that their rights are respected consistent with the provisions of their respective national laws or, in the absence of national legislation, in accordance with the UN Declaration on the Rights of Indigenous Peoples.

To achieve both primary objectives (carbon emission reduction) and secondary objectives (biodiversity conservation), REDD initiatives can benefit from prior experiences with successful protection of standing forest resources. Estrada Porría, Corbera, and Brown (2007) review the effectiveness of existing policy and legal instruments to halt deforestation, concluding that enhanced security of tenure seems to be the most effective approach. Such policies apply to forest dwellers as well as to long-term logging concessionaires; they can reduce perverse incentives to clear cut while also creating incentives to implement sustainable forest management. Their effectiveness depends in part, though, on political changes and on respect for and enforcement of legal rights.

As well as helping with carbon sequestration, indigenous peoples can play an important role in protecting biodiversity. Because a number of important biodiversity hot spots in LAC are not protected legally through protected-area instruments, and indigenous peoples live in some of these hot spots, it is important to integrate these peoples and their rights into schemes for the protection of biodiversity. Of the world's remaining biodiversity-rich areas, a large proportion—80 percent, according to Adamson and Tawake (2007)—are also indigenous peoples' homelands. Based on this recognition, important steps have been taken, such as the protection of 2 million km^2 of tropical forest areas for indigenous peoples and biodiversity, half of which is in the Brazilian Amazon (Pimm and others 2001). However, important biodiversity is still under pressure and enjoys no legal or political protection, and much of the officially protected land is far from being sustainably managed.

For the Amazon basin, Soares-Filho and others (2006) estimate that by 2050, the cumulative avoided deforestation potential could reach 62,000 metric tons of CO_2 under a "governance" scenario (IPCC 2007) in which Brazilian environmental legislation is implemented across the Amazon basin through the refinement and multiplication of experiments regarding the "enforcement of mandatory forest reserves on private properties through a satellite-based licensing system, agro-ecological zoning of land use, and the expansion of the PA network [Amazon Region Protected Areas Program]." Added to this list should be the enforcement of indigenous peoples' rights to standing forests by establishing clear property rights over land and paying its owners not to cut down trees.

In many developing countries, community conservation areas have emerged locally as part of customary resource-management and sustainable-use practices. Often these areas are locally governed, without formal recognition or formal legal backing. Thus, if conflicts arise over the use or right of access to such areas and their resources, systems of customary law and practice are applied to resolve the problems. However, legal disputes may ensue when statutory laws clash with these customary means of conflict resolution. Such legal wrangles are not necessarily detrimental to conservation; they may offer valuable lessons about how to integrate customary and statutory laws, and formal and nonformal institutions, for more effective conservation. Community conservation areas are conducive to sustainable biodiversity protection and—if legally enshrined— are enforceable. In indigenous territories that are governed by customary law, community-conserved areas express indigenous peoples' rights as enforceable claims at the local community level and part of enforceable rules that govern intercommunity behavior. Recent evidence, covering the entire Brazilian Amazon, shows the potential benefits of this approach (Nepstad and others 2006). In LAC, the demarcation of indigenous lands and territories legally establishes the right of indigenous peoples to rule by customary law within these territories. Such demarcation has tended to occur in direct response to threats from commercial extractive interests, urbanization, and migrating settlers, but with changing demands and social changes there are huge needs for support to facilitate the continued protection of forest resources.

Currently there are few experiences with REDD-like schemes, including private REDD schemes or payment for environmental services. One such scheme is in Brazil's Amazonas state, where US$ 8.1 million from private companies such as Marriott Hotels and Bradesco, a large bank, is being handed over by the state government to 6,000 families in exchange for their not cutting down any more trees. The challenge is to extend such schemes to the trees on the edge of the farmland, which are most at risk.

In Brazil, as the network of protected areas expands from 36 to 41 percent of the Brazilian Amazon over the coming years, the greatest challenge will be to establish successful reserves and strengthen indigenous people's rights to land in areas at high risk from frontier expansion. Success will depend on broad political backing. This backing would rely mainly on two arguments: First, indigenous peoples' management of biodiversity causes no harm, or less harm to biodiversity than other types of management, because of their special interests in safeguarding it. Second, biodiversity is better protected and cared for by indigenous peoples as a result of their culturally developed livelihood strategies and unique indigenous knowledge of the management of biodiversity.

Both these arguments situate indigenous peoples as partners in the protection of biodiversity and are mutually linked, but they require different types of support. The first requires political, legal, and institutional support to uphold indigenous peoples' rights to territory, land, and resources, including through protection and monitoring of physical borders and providing support to withstand external and internal agents of pressure such as timber companies and local corruption. This support needs to take into account the national context, including the roles of state-exerted support and pressure, private sector support and pressure (from NGOs, firms, neighboring landowners), and indigenous peoples' organizations. The second argument calls for strengthening the conditions for continued development and maintenance of indigenous knowledge regarding conservation and use of biodiversity. Because knowledge is generated in indigenous peoples' institutions and practices, strengthening these peoples' knowledge on conservation and biodiversity will entail supporting the necessary conditions for continued development of their livelihood strategies (Kronik 2001). Support should also continue for sustainable income generation to ease indigenous people's dependence on actors seeking short-term gains from extracting resources that are important for long-term use and conservation of biodiversity.

The Amazon basin provides a good example of the point that threats from climate change and variability should not be seen in isolation, but in relation to other threats, such as those from expansion of the agricultural frontier, advancing colonization, political unrest, and illegal coca cultivation—all of which put pressure on the authority of traditional cultural institutions. The effect of these trends on local governance appears to be significant, while the presence of the state in many of the areas inhabited by indigenous peoples is weak. Better governance is important for achieving successful adaptation and for implementing and adopting mitigation instruments—underlining the need to strengthen institutional capacities. There is a need to focus on increasing the capacity of governments to formulate and implement sound policies, enabling the exercise of political authority at all relevant levels. Efforts for this purpose must think in terms of indigenous peoples' needs, resources, and institutions and must incorporate these into the broader development framework.

The success of national-scale REDD initiatives will depend not only on strengthening developing countries' technical and institutional capacities, but also on addressing weaknesses in governance. When indigenous peoples' organizations call for alternatives to the commoditization and "carbonization" of the environment they live in and depend on, such alternatives imply a lengthy effort to strengthen rights and improve governance in the countries they live in. This effort will need to go beyond the climate-change and even the environmental agendas—making it difficult to foresee massive emissions reductions from avoided deforestation in the short term.

More generally, it will be difficult to achieve either climate-change adaptation or mitigation without strengthening the necessary conditions for continued use and development of indigenous peoples' knowledge. This calls for a dialogue between knowledge systems. For example, there is an urgent need to support and develop monitoring of the changing climate. Larger monitoring systems are needed that provide reliable, responsive, and relevant local information, but so, too, are local monitoring and forecast systems, combined with access to local knowledge on how to interpret changes and how the changes may affect local livelihood strategies. Combining knowledge of different types raises important issues about information taxonomy, communication methods, and how to deal with uncertain data. Indigenous peoples' knowledge is usually treated largely as ahistorical and timeless data that can be merged into current plans and programs. However, this approach does not provide opportunities for understanding specific political and historical conditions for developing, maintaining, and disseminating this knowledge. To craft climate-change adaptation responses and mitigation efforts, the negotiation, design, and implementation of local and global knowledge systems must be participatory so that relevant processes, institutions, and practices are protected, consulted, and included.

Notes

1. In the high Andes, the ablution and accumulation seasons coincide, so that glaciers are the main seasonal water reserve; this makes the rise in temperature a bigger problem than in midlatitude mountains (Vuille and others 2008).

2. Migration patterns also reflect political violence in both Chiapas and Guatemala.

3. Manioc, the staple crop in Indian gardens in this subregion, is relatively resistant to heat and drought and grows well even in poor soils. Though it guarantees a source of carbohydrates, it leaves important nutritional needs unmet.

4. Including the milpa, home gardens, secondary forest, aquatic systems, and old-growth forest.

5. "From satellite-based maps of land cover and fire occurrence in the Brazilian Amazon, the performance of large uninhabited parks (>10,000 ha) and inhabited (indigenous lands, extractive reserves, and national forests) reserves were compared... No strong difference was found between parks and indigenous peoples' lands. However, uninhabited reserves tended to be located away from areas of high deforestation and burning rates, while in contrast, indigenous lands were often created in response to frontier expansion, and many prevented deforestation completely, despite high rates of deforestation along their boundaries." (Nepstad and others 2006)

6. The World Bank is studying the forest sector and REDD. It would seem a natural follow-up to this volume to initiate a similar piece on the social impacts associated with climate-change mitigation in forestry.

7. Specifically, the Anchorage Declaration recommends that the United Nations Framework Convention on Climate Change (UNFCCC) organize regular technical briefings by indigenous peoples on traditional knowledge and climate change; recognize and engage the International Indigenous Peoples' Forum on Climate Change and its regional focal points in an advisory role; immediately establish an indigenous focal point in the secretariat of the UNFCCC; appoint indigenous peoples' representatives in UNFCCC funding mechanisms in consultation with indigenous peoples; and take the necessary measures to ensure the full and effective participation of indigenous and local communities in formulating, implementing, and monitoring activities, mitigation, and adaptation to impacts of climate change.

8. Paragraphs 106–110.

References

Aguayo, Eva, Pilar Exposito, and Lawelas Nelida. 2001. "Econometric Model of Service Sector Development and Impact of Tourism in Latin American Countries." *Applied Econometrics and International Development* 1(2).

Allison, Edward H., W. Neil Adger, Marie-Caroline Badjeck, Katrina Brown, Decian Conway, Nick K. Dulvy, Ashley Halls, Allison Perry, and John D. Reynolds. 2005. *Effects of Climate Change on the Sustainability of Capture and Enhancement Fisheries Important to the Poor: Analysis of the Vulnerability and Adaptability of Fisherfolk Living in Poverty.* Fisheries Management Science Program: Final Technical Report. London: U.K. Department for International Development.

Allison, E.H., et al. (2009). "Vulnerability of national economies to the impacts of climate change on fisheries." *Fish and Fisheries* 10: 173–96.

Anderson, James J. 2000. "Decadal Climate Cycles and Declining Columbia River Salmon." In *Sustainable Fisheries Management: Pacific Salmon*, ed. E. E. Knudsen and Donald MacDonald, 467–84. New York: CRC Press.

Barros, V. 2005. "Global Climate Change and the Coastal Areas of the Río de la Plata." Final Report for Project No. LA 26. Submitted to Assessments of Impacts and Adaptation to Climate Change. http://www.aiaccproject.org/.

Becken, Suanne and John Hay 2007. *Tourism and Climate Change: Risk and Opportunities*. Channel View Publications. UK, Clevedon; USA, Tonawanda; Canada, North York.

Bueno, Ramón, Cornelia Herzfeld, Elizabeth A. Stanton, and Frank Ackerman. 2008. "The Caribbean and Climate Change: The Costs of Inaction." Tufts University. Available at http://ase.tufts.edu/gdae/Pubs/rp/Caribbean-full-Eng.pdf.

Burke, L., and J. Maidens. 2004. *Reefs at Risk*. World Resources Institute (WRI). Washington, DC.

Carrol, Rory. 2008. "Tourism Curbed in Bid to Save Galapagos Haven," *Mail and Guardian*, October 12. http://www.mg.co.za/article/2008-10-12-tourism-curbed-in-bid-to-save-galapagos-haven.

Cavallos, Diego. 2005. "Untreated Wastewater Making the Sea Sick." Interpress Service News Agency.

Center for Remote Sensing of Ice Sheets (CReSIS). 2008. Sea Level Rise Maps and GIS Data – Southeast USA. https://www.cresis.Ku.edu.

CHA-CTO (Caribbean Hotel Association and Caribbean Tourism Organization). 2007. "CHA-CTO Position Paper on Global Climate Change and the Caribbean Tourism Industry." http://caribbeanhotelassociation.com/downloads/Pubs_ClimateChange0307.pdf .

De la Torre, Augusto, Pablo Fajnzylber, and John Nash. 2009. *Low Carbon High Growth: Latin American Responses to Climate Change. An Overview*. World Bank Latin American and Caribbean Studies. Washington, DC: World Bank.

DFID (Department for International Development, U.K.). 2004. *Climate Change in Latin America*. Key Fact Sheet No. 12. Key Fact Sheets on Climate Change and Poverty. http://www.dfid.gov.uk/pubs/files/climatechange/keysheetsindex.asp.

Drosdoff, Daniel. 2009. "Barbados acts to prevent water crisis." *Newsbeat, IDB America: Magazine of the Interamerican Development Bank*. http://www.iadb.org/idbamerica/index.cfm?thisid=2793

Ehmer, Philippe, and Eric Heymann. 2008. *Climate Change and Tourism: Where Will the Journey Lead?* Deutsche Bank Research, Frankfurt am Main.

Fay, Marianna, ed. 2005. *The Urban Poor in Latin America.* A World Bank Report. Washington, DC: World Bank.

Freeman, Gary E., and Craig Fischenich. 2000. "Gabions for Streambank Erosion Control." U.S. Army Corps of Engineers, Environmental Laboratory. http://el.erdc.usace.army.mil/elpubs/pdf/sr22.pdf.

FAO (Food and Agricultural Organization). 2007. Fisheries and Aquaculture Information and Statistics Service. FAO.

———. 2008. Fisheries and Aquaculture Country Profiles. http://www.fao .org/fishery/countryprofiles/search/en.

Hall, Michael C. and James Higham (eds). 2005. *Tourism, Recreation, and Climate Change.* Bristol, UK: Channel View Publications.

Hamilton, Jacqueline M., D. J. Maddison., and R. S. J. Tol. 2005. "Climate Change and International Tourism: A Simulation Study." *Global Environmental Change* 15: 253–66.

Hendershot, Rick. 2006. "Cancun Beach Being Restored at Record Pace. Health Guidance for Better Health." http://www.healthguidance.org.

Hinrichson, Don. 2008. "Ocean Planet in Decline." *People and Planet: People and Coasts and Oceans.* http://www.peopleandplanet.net/doc.php?id=429& section=6.

Hoeg-Guldberg, O., P. J. Mumby, A. J Hooten, R. S. Steneck, P. Greenfield, E. Gomez, C. D. Harvell, P. F. Sale, A. J. Edwards, K. Caldeira, N. Knowlton, C. M. Eakin, R. Iglesias-Prieto, N. Muthiga, R. H. Bradbury, A. Dubi, and M. Hatsioloz. 2007. "Coral Reefs under Rapid Climate Change and Ocean Acidification." *Science* 318 (December 14).

International Labor Organization (ILO). 2008. LABORSTA – Database of Labor Statistics. Statistics and databases. http://laborsta.ilo.org.

International Union for Conservation of Nature (IUCN). 2007. "Latin American Park Congress: 2008–2018 to be 'decade of MPAs.' http://www.iucn.org.

Jackson, Ivor. n.d. "Potential Impacts of Climate Change on Tourism." Issues paper prepared for OAS – Mainstreaming Adaptation to Climate Change (MACC) Project.

Kawasaki, T. 2001. "Global Warming Could Have a Tremendous Effect on World Fisheries Production." In *Microbehavior and Macroresults: Proceedings of the Tenth Biennial Conference of the International Institute of Fishery Economics and Trade,* ed. R. S. Johnston and A. L. Shriver. July 10–15, 2000, Oregon.

Kelly, Kathryn, and Ronald Sanabria. 2008. "Mainstreaming Biodiversity Conservation into Tourism through the Development and Dissemination of Best Practices." World Bank presentation, September 18, World Bank, Washington, DC.

Latin American and Caribbean Region Social Development (LCSSD). 2008. "Tourism in LCSSD: Building a New Beam." A World Bank Presentation.

Lydersen, Kari. 2008. "Risk of Diseases Rises with Water Temperatures." *Washington Post*, October 20, A08.

Martinez, Julia. 2008. "Impacts of Climate Change in the Tourism Sector in Mexico." World Bank presentation, April 15, World Bank, Washington, DC.

McField, Melanie, and Patricia Kramer. 2007. *Healthy Reefs for Healthy People: A Guide to Indicators of Reef Health and Social Well-Being in the Mesoamerican Reef Region*. Washington, DC: Smithsonian Institution.

McGoodwin, James R. 2001. "Understanding the Cultures of Fishing Communities: A Key to Fisheries Management and Food Security." Rome: FAO. http://www.fao.org/docrep/004/y1290e/y1290e00.HTM.

McGranahan, Gordon, Deborah Balk, and Bridget Anderson. 2007. "The Rising Tide: Assessing the Risks of Climate Change and Human Settlements in Low-Elevation Coastal Zones." *Environment and Urbanization* 19 (1): 17–37.

McLeod, Elizabeth, and Rodney V. Salm. 2006. *Managing Mangroves for Resilience to Climate Change*. Gland, Switzerland: World Conservation Union.

Morrison, Jason, Mari Morikawa, Michael Murphy, and Peter Schulte. February 2009. *Water Scarcity and Climate Change: Growing Risks for Business and Investors*. A CERES and Pacific Institute Report.

Murray, Peter A. n.d. *Climate Change, Marine Ecosystems, and Fisheries: Some Possible Interactions in the Eastern Caribbean*. Environment and Sustainable Development Unit, Organization of Eastern Caribbean States.

Nellemann, C., and E. Corcoran, eds. 2006. *Our Precious Coasts-Marine Pollution, Climate Change, and the Resilience of Coastal Ecosystems*. Grid-Arendal, Norway: UNEP.

OECD (Organization for Economic Cooperation and Development). 2007. *Ranking Port Cities with High Exposure and Vulnerability to Climate Extremes: Exposure Estimates*. OECD Environment Directorate. Paris: OECD.

Perry, Allison L., et al. "Climate Change and Distribution Shifts in Marine Fishes." *Science Magazine* 308 (June 24, 2005).

Ravallion, Martin, Shaohua Chen, and Prem Sangraula. 2008. *New Evidence on the Urbanization of Global Poverty*. Background Paper for the World Development Report 2008. Development Research Group, World Bank.

Sale, P. F., M. J. Butler IV, A. J. Hooten, J. P. Kritzer, K. C. Lindeman, Y. J. Sadovy de Metcheson, R. S. Steneck, and H. van Lavieren. 2008. *Stemming Decline of the Coastal Ocean: Rethinking Environmental Management*. Hamilton, Canada: United Nations University, UNU-UNWEH.

Sumaila, Ussif Rashid, Sylvie Guenetta, Jackie Alder, David Pollard, and Ratana Chuenpagdee. 1999. *Marine Protected Areas and Managing Fished Ecosystems.* CMI Reports. Bergen: Chr. Michelsens Institute.

Tropical Re-Leaf Foundation. 2008. "Nariva Swamp Restoration Initiative." http://www.ema.co.tt/main.htm.

United Nations Environment Programme (UNEP). 2007. Climate Change Hits Hard on Latin America and the Caribbean. UNEP Press Release. http://www.unep.org.

UN Habitat (United Nations Human Settlement Program). 2003. *The Challenge of Slums. Global Report on Human Settlements.* London: Earthscan Publications.

United Nations Population Division. 2006. *World Urbanization Prospects: The 2005 Revision.* Department of Economic and Social Affairs, Urban and Rural Areas Dataset (POP/DB/WUP/Rev.2005/1/table A.6), dataset in digital form.

UNWTO (United Nations World Tourism Organization). 2008. "Climate Change and Tourism: Responding to Global Challenges." Advanced Summary. http://www.unwto.org/index.php.

Vergara, Walter. 2005. *Adapting to Climate Change: Lessons Learned, Work in Progress, and Proposed Next Steps for the World Bank in Latin America.* Sustainable Development Working Paper 25. World Bank, Washington, DC.

Vergara, Walter, Natsuko Toba, Daniel Mira-Salawa and Alejandro Deeb. 2009. The consequences of climate-induced coral loss in the Caribbean by 2050–2080. In *Assessing the Potential Consequences of Climate Destablization in Latin America.* Sustainable Development Working Paper. Washington, DC: World Bank.

Verner, Dorte, and Willy Egset. 2008. *Social Resilience and State of Fragility in Haiti.* A World Bank Report. Washington, DC: World Bank.

WHO (World Health Organization). 2008. *Global Epidemics and Impacts on Cholera.* http://www.who.int/topics/cholera/impact/en/index.html.

Wilkinson, C. R., ed. 2008. *Status of Coral Reefs of the World: 2008.* Townsville, Australia: Global Coral Reef Monitoring Network and Rainforest Research Center.

Wilkinson, Clive, and David Souter, eds. 2008. *Status of the Caribbean Coral Reefs After Bleaching and Hurricanes in 2005.* Townsville, Australia: Global Coral Reef Monitoring Network.

World Atlas of Coral Reefs, 2001. Revised 2007. UNEP/WCM. http://www.unep-wcmc.org/marine/coralatlas/index.htm.

World Bank. 2001. "Attacking Brazil's Poverty: A Poverty Report with a Focus on Urban Poverty Reduction Policies." Report No. 20475-BR, Washington, DC: World Bank.

————. 2007. "Sustainable Fisheries through Improved Management and Policies." Chapter 6 in *Republic of Peru Environmental Sustainability: A Key to Poverty Reduction in Peru.* Environmentally and Socially Sustainable Development in Latin America and the Caribbean. Washington, DC: World Bank.

————. 2008. "At a Glance: Poverty Numbers in LAC." http://www.worldbank .org/lacpoverty.

WRM (World Rainforest Movement). 2002. Latin America: Mangroves: Local Livelihoods vs. Corporate Profits. December.

WTTC (World Travel and Tourism Council). 2005. "Global Travel and Tourism Poised for Continued Growth in 2005: Tsunami Impact on Travel and Tourism is Significant but Limited." *Tourism News.* http://www.wttc.org/eng/ Tourism_News/Press_Releases/Press_Releases_2005.

CHAPTER 6

Conclusions and Recommendations

Indigenous peoples across Latin America already perceive and suffer a range of effects of climate change and variability. These effects compound other important stresses on their productive resources and traditional ways of life, threatening the viability of their livelihoods and eroding their traditional cultures. As existing climate change trends are projected to intensify, with disproportionate effects on the most poor and vulnerable, it becomes increasingly urgent to cooperate with indigenous groups and draw on their knowledge of the local context to identify and implement environmentally and culturally sustainable adaptation measures.

Cultural, social, and biophysical differences among indigenous peoples and the places they inhabit matter. Climate change affects them in different ways, and it is important to consider these differences. For rural indigenous people across the LAC region, our findings show that increased water scarcity, rising mean temperatures, and disturbances in seasonal rhythms are affecting the viability of crop and livestock production and the availability of food foraged from the wild. Often the effects of climate change result in food insecurity and poor health, as well as an erosion of confidence in the solutions provided by traditional cultural institutions and authorities.

Indigenous peoples' capacity to adapt to climate change events has often already been stretched to the limit. Because environmental changes are happening so fast, indigenous communities find it difficult to adapt in a culturally sustainable manner. Not only is their livelihood threatened, but so is their cultural integrity. Social and cultural cohesion is vital to their survival as indigenous people, and rising concerns about food security and poor health prompt the younger generation and even whole families to pursue new adaptation measures such as temporary and permanent migration, adoption of modern agricultural methods, and income generation from new sources such as tourism. These adaptation schemes leave the traditional way of life behind. Most of the thousands of indigenous people who migrate to cities crowd into poor, urban neighborhoods.

Indigenous peoples' knowledge of the local environment is key to finding solutions that will sustain their way of life, while helping them meet demand from millions of people who depend on them as agricultural producers. Climate-change adaptation and mitigation strategies implemented by government agencies and NGOs can achieve viable outcomes for indigenous cultures and livelihoods only if they make a concerted effort to incorporate local knowledge systems and institutions and to be culturally, environmentally, and economically appropriate.

This study is based on fieldwork in five case-study countries in three distinct ecological regions—the Amazon, the Andes and sub-Andes, and the Caribbean and Mesoamerica—that were selected based on the main types of climate changes besetting LAC. In Bolivia, Colombia, Mexico, Nicaragua, and Peru, we analyzed how diverse indigenous groups perceive and respond to the effects of climate change. We used methods such as key informant interviews, participatory climate-change-impact scenarios, and institutional analyses.

Indigenous Peoples of Latin America and the Caribbean

LAC has nearly 40 million indigenous people belonging to more than 600 different ethno-linguistic groups, each with a distinct language and world view. The majority of indigenous people live in the colder and temperate high Andes and in Mesoamerica. However, the majority of populations with distinct ethno-linguistic identities are found in the warm tropical lowlands, most notably in the Amazon rain forest. While several indigenous cultures survive in urban settings, this study focuses on rural indigenous populations.

Most of these populations live in extreme poverty, generally having little formal education, few productive resources, few work skills applicable in the market economy, and limited political voice. Like other poor people who depend on natural resources, their livelihoods are vulnerable to the effects of climate change.

But what sets indigenous peoples apart from nonindigenous people, and makes them especially vulnerable, is the intimate ways in which they use and live off natural resources and their dependence on cultural cohesion. Their livelihood strategies are maintained through social and cultural institutions, practices, and knowledge systems that are based on experiments with nature, juxtaposed with a stock of knowledge developed over time and passed on through generations. Their ability to predict and interpret natural phenomena, including weather conditions, has not only been vital for their survival and well-being, it has also been instrumental in the development of their cultural practices, social structures, trust, and authority.

Culture, Livelihood, Institutions, and Knowledge

Indigenous peoples' cultural institutions strongly affect their natural resource management, health, and coping abilities. The annual succession of seasons is important for them: it orders the timing of the agricultural cycle and of ritual practices that help them to prevent illnesses and promote well-being, and it is crucial for the reproduction of wildlife. The interplay between nature's cycles and certain cultural practices leads to the creation of cultural capital, which in turn is reproduced through regularized practices and rituals. Cultural institutions are developed around these practices and rituals, and they serve to maintain, develop, and test information and thereby contribute to the social generation of knowledge.

Basic to many indigenous peoples' understanding of the relationship between society and nature is the notion of maintaining balance between the human, natural, and cosmological realms, based on trusted social and cultural knowledge and practices.[1] These balances are constantly in flux, and living and acting involves negotiating them. So when changes occur, for example, in climatic conditions, people look to themselves and their social institutions and rituals as possible reasons. If they cannot amend the way they conduct their own lives, they seek other means to restore balance. The cultural rituals and social corrective measures that are used vary from place to place, but they share the striving for balance between the social and the natural, based on trusted social and cultural knowledge and practices.

Traditionally, the viability of indigenous peoples' livelihoods has depended heavily on the vitality of their knowledge systems developed over generations. But increasingly, the traditional knowledge and practices are obsolete. In particular, they are increasingly at odds with rainfall patterns. Key members of indigenous communities, interviewed in different areas, said that the natural signs and indexes they now perceive are alarming, including growing scarcity of water; erosion of ecosystem and natural resources, for example, through salinization of soils; changes in biodiversity as a consequence of the spread of alien species; plant diseases affecting crops; a higher death toll among livestock; higher risks of infectious diseases; and crop losses. Seasonal variation has become so unpredictable that the adaptation strategies developed to tackle the normal span of variation no longer provide the necessary security. Productivity has been falling, and maintaining a livelihood has become more difficult, as a result of—for example—water shortages for crops and livestock, declining crop yields, unfamiliar pests, hailstorms, and unseasonal frosts. Changes in precipitation and seasonal regimes not only disrupt the agricultural calendar, but affect the availability of river fish, wild fruit, and game. Specific sources of vulnerability to the threats posed by climate change differ widely among indigenous peoples, influenced by access to land, resources, types of livelihood strategies, and institutions.

When climatic conditions become impossible to predict, the elders and traditional leaders—who are the experts within traditional knowledge systems—lose credibility. When they can no longer guarantee abundance and prosperity through applying their traditional knowledge and rituals, their status falls. Consequences follow for their social organizing capacity and the maintenance of culture, and societies become "rudderless." People look elsewhere for solutions to their problems, by both turning to other bodies of knowledge and migrating. Across the region, interviewees from several different groups of indigenous peoples referred to this disintegrative process as leading to the end of their life as indigenous peoples. Not only do individual family members leave, but whole families are now uprooting (box 6.1). Some communities are left ghostlike; in others, only young children and the elderly remain. This phenomenon is seen in communities across the region, from Argentina to Mexico. The induced migration toward the cities, where people crowd into poor, conflict-ridden neighborhoods on urban fringes, creates the need for government solutions (Verner 2010). A much larger-scale parallel process of partly planned extensive migration to the lowlands draws on earlier traditions, but it too has consequences for vulnerability to climate change and variability. There

---

Box 6.1

Indigenous Peoples and Migration Related to Climate Change

Only cursory estimates exist of the level of climate-change-induced migration among LAC's indigenous peoples. And very few empirical studies have examined the nexus between climate change and migration among these peoples. The migration drivers for indigenous peoples, like other vulnerable groups, are diverse and linked to access to economic, cultural, and social capital. Across LAC, internal migration has gradually become essential to subsistence for many rural indigenous peoples because of their lack of access to land, low productivity, and persistent poverty. In impoverished indigenous communities, men migrate seasonally to work on construction or other nonskilled jobs, leaving women in charge of children and domestic work.

Source: Perch-Nielsen, Bättig, and Imboden (2008).

---

may be more water in the lowlands, but people there are often more dependent on the market economy.

Lack of an institutional framework in most Latin American countries to facilitate a dialogue between indigenous peoples' organizations and public authorities at both the national and regional levels prevents the effective participation of indigenous peoples in climate-change adaptation initiatives. Consequently, most public institutions and municipal authorities lack knowledge of indigenous peoples' life, culture, and needs, impairing their relationship with local communities when public adaptation programs do not adequately consider the specific climate-change concerns of indigenous peoples.

Impact of Climate Change on Indigenous Peoples

In the Andean and sub-Andean regions, the climatic changes translate into increasing mean temperatures—causing drought and glacier retreat—and variations in seasonality—marked by changes in the patterns and intensity of rainfall, hailstorms, and frosts. The changes affect the Aymara and Quechua peoples by putting food security at risk and affecting social stability, health, and psychological well-being. High-elevation Andean herders' livelihoods are threatened by the melting of the glaciers and resulting water scarcity in dry periods. Also particular to the high mountains are the effects of unexpected hailstorms and frosts during the crop-growing season. In the lower Andes, the melting of glaciers causes floods and soil erosion.

In the Caribbean and Mesoamerica, the increased intensity of extreme events, notably hurricanes, endangers entire ecosystems—as well as directly endangering people's lives—while gradual warming and acidification of the oceans threaten the viability of the mangroves and coral reefs that some communities rely on for their livelihoods. Indigenous peoples affected by climate change and variability in this part of LAC include the Miskitu, the Sumu-Mayangna complex (Ulwas, Twaskas, and Panamaskas), and the Rama. The climate-change projections are for more powerful hurricanes and a continuing upward trend for intense precipitation followed by extended periods of drought stemming from higher temperatures. Indigenous people interviewed in Nicaragua said that even 20 years after Hurricane Joan, they have yet to fully recover the abundance of forest resources (especially lumber and wildlife) that existed before the hurricane. Part of the reason is that the agricultural frontier expanded after the hurricane felled large tracts of forest and thus gave settlers easy access to colonize indigenous territories. In contrast, fish catches and agricultural yields have regained their prehurricane levels—after five years for fishing and perennial crops such as fruit trees, and after just one year for rice, beans, banana, plantain, cassava, and maize.

In the lowland forests of the Amazon region, indigenous people's livelihoods are more affected by changes in precipitation and seasonality than by increasing temperature. Often crops now fail repeatedly, limiting the diversity of sources of nutrition and the intake of needed nutrients. River fish and turtles are important sources of food for these communities, but their reproduction has been badly damaged because the rivers no longer rise and fall seasonally, as they used to. The greatest concern of indigenous peoples in the Amazon region, however, is their social situation. The traditional harmony of their lives with nature is disturbed not only by climate change and variability, but also by the effects of advancing colonization, destruction of the forest, political unrest, illegal coca cultivation, excessive resource exploitation, gold mining, and trade. In combination, these changes lead to increasing social disarray.

In all of the regions we studied, human health is perceived as a key area of impact. Particular concerns are the spread of disease vectors into places where they could not previously thrive (in Andean regions), increased incidence of respiratory and diarrheal diseases (in the Amazon region), and widespread increased difficulty in obtaining adequate nutrition. In turn, malnutrition has consequences for people's ability to resist infectious diseases, and it compromises the development prospects of children who survive it.

Climate-Change Adaptation in an Indigenous Context

As regards efforts to adapt to the effects of climate change, and to mitigate climate change itself, the research reported in this study brings out five important messages:

- Climate-change adaptation and mitigation efforts take place within a context made up of social, economic, and natural forces and actors.
- Social, political, cultural, and environmental forces determine exposure and sensitivity to the effects of climate change and shape adaptive capacity locally.
- Larger processes and instruments such as rights, laws, and economic drivers are at least as important to address as local-level adaptation and mitigation efforts.
- No adaptation effort should be started for climate-change purposes alone. It is imperative to mainstream adaptation into overall development policies, plans, and programs.
- Cultural values that have developed over time in indigenous peoples' institutions to deal with uncertainty and variability may be the main contribution in developing adaptation and mitigation strategies.

Across the continent, indigenous peoples are adjusting their productive activities to try to address the effects of climate change and other new influences. Some communities have shown spontaneous and intuitive capacity to adapt, helped by their cultural institutions, which shape the management of natural resources. In some cases, experiences with training projects, as well as technical assistance in agriculture and other areas, have indirectly helped indigenous people to identify alternatives and hence increase their adaptive capacity. By contrast, some of the other communities visited are failing to produce enough food and must change their livelihoods so drastically that they lose vital conditions for the development and reproduction of their culture.

Given the high human costs already being felt, it is vital that further strategies be developed to help indigenous communities adapt to climate change and variability. Recommendations for this purpose are offered at the end of this chapter. The fact that many indigenous people live in an institutional vacuum, without any linking social capital to institutions outside their own society, underlines the importance of taking the political and governance dimensions of their adaptation into account.

Participatory adaptation is essential to achieving sustainable solutions. Indigenous peoples' knowledge is often treated as ahistorical and timeless data that can be merged into current plans and programs. However, this approach does not provide opportunities for understanding the specific political and historical conditions for the development, maintenance, and dissemination of indigenous knowledge. To craft adaptation responses to local and global knowledge systems, indigenous peoples must participate in the negotiation, design, and implementation of solutions so that relevant processes, institutions, and practices are protected.

Climate-Change Mitigation in an Indigenous Context

As Latin American countries develop and negotiate efforts to mitigate climate change, indigenous communities can make important contributions by acting as stewards of natural resources and biodiversity in the territories they live in, provided their rights are recognized and respected. Recent studies show that areas governed by indigenous peoples are less prone to deforestation than other areas.

In LAC, where the burning of forests is one of the main sources of greenhouse gas, several countries are proposing launching programs to reduce emissions from deforestation and degradation (REDD).[2] In the LAC context, it is difficult to imagine much REDD without indigenous peoples' participation, simply because they control and often own large tracts of dense forest. If forest protection is to be an effective climate-change mitigation mechanism in LAC, indigenous peoples' rights must be recognized when designing and negotiating agreements for this purpose. Indigenous peoples fear having their autonomy and authority undermined by entering into government-negotiated REDD agreements.

As well as helping with carbon sequestration, indigenous peoples can play an important role in protecting biodiversity. Indigenous lands occupy one-fifth of the Brazilian Amazon—five times the area under protection in parks—and are currently the most important barrier to Amazon deforestation. As the protected area network expands in coming years to include 36–41 percent of the Brazilian Amazon, the greatest challenge will be successful reserve implementation in areas at high risk of frontier expansion. Success in maintaining forest reserves will depend on broadly based political backing.

This backing would rely mainly on two arguments: First, indigenous peoples' management of biodiversity causes no harm, or less harm to biodiversity than other types of management, because of their special interests in safeguarding it. Second, biodiversity is better protected and cared for by indigenous peoples because of their culturally developed livelihood strategies and unique indigenous knowledge of the management of biodiversity.

The first argument requires political, legal, and institutional support to uphold indigenous peoples' rights to territory, land, and resources—including through protection and monitoring of physical borders—and support to withstand external and internal agents of pressure such as timber companies and local corruption. This support needs to take into account the national and local context, including the roles of state-exerted support and pressure, private sector support and pressure (from NGOs, firms, neighboring landowners), and indigenous peoples' organizations.

The second argument calls for strengthening the conditions for continued development and maintenance of indigenous knowledge regarding the conservation and use of biodiversity. Because knowledge generation takes place in various indigenous peoples' institutions and practices, strengthening indigenous peoples' knowledge of conservation and biodiversity will entail supporting the necessary conditions for continued development of their own livelihood strategies (Kronik 2010). Support should also continue for sustainable income generation to ease indigenous people's dependence on actors seeking short-term gains from extracting resources that are important for long-term use and conservation of biodiversity.

Operational Recommendations

Large proportions of the most vulnerable indigenous people lack adequate coping strategies of their own. Climate-change adaptation strategies and mitigation instruments are needed that will strengthen the capacity of indigenous peoples to (a) use existing resources to increase their resilience, as they adapt production systems to stabilize and increase their food security; (b) resolve conflicts over the use of resources; and (c) identify and advocate policy alternatives that enhance adaptive capacity.

Agriculture (cropping and pastoralism) is by far the most important source of income for rural indigenous people in the region; hence, the susceptibility of agriculture to climate change and variability directly affects their livelihoods. It is vital that adaptation strategies addressing this vulnerability are effective.

Governments may help vulnerable indigenous communities increase their social resilience to climate change and variability by ensuring that their informal institutions (as distinct from their formal, political organizations) play a role in community-level negotiations of government-promoted adaptation and mitigation initiatives. This will be particularly important in addressing the use and conservation of natural resources, containing deforestation, and upholding traditional indigenous peoples' cultural institutions and livelihoods.

In a two-pronged approach, the following recommendations pertain to the adaptation and disaster preparedness intended to reduce the negative social impacts on indigenous peoples of both extreme events, like hurricanes, and more subtle changes, for example, in precipitation regimes, access to water, and the length of droughts and frost-free periods.

Adaptation measures should focus equally on processes and outcomes, applying participatory strategies to build resilience:

- Strengthen strategies and increase support to help indigenous communities adapt to climate change and variability, using a participatory process that acknowledges the diversity of their livelihood strategies and the role and efficiency of their cultural institutions. This support must be rooted in, or at a minimum must respect, the traditional knowledge of indigenous peoples.
- Extend technology and technical advice on agricultural practices that are culturally, environmentally, and economically appropriate.
- Promote the protection of indigenous peoples' rights to land and natural resources, including models of collective ownership.
- Include representatives from indigenous peoples' organizations in both national and local climate-change adaptation initiatives. Participatory processes should be implemented at the design stage and followed through to the final project evaluation. As in all planning and implementation processes, a concerted, integrated effort is needed to avoid "business as usual" or simply relabeling ongoing operations.
- Strengthen indigenous peoples' organizations, and enhance institutional dialogue with national and regional public authorities.
- Increase access to conditional cash transfers and other government programs to increase the resilience of indigenous peoples to the effects of climate change.

To build disaster preparedness, the focus must be on strengthening institutions, building awareness, and implementing legal requirements protecting local territories:

- Establish and strengthen early-warning systems for hurricanes, and invite the participation of indigenous peoples and poor farmers (for example, a network of radio transmitters to prepare communities for climate-related events in coordination with community radio broadcasting). Establish clear guidelines on which languages to use for broadcasting the messages. After a hurricane or similar disaster, agencies undertaking postdisaster management should seek the participation of indigenous people to facilitate the recovery of local communities and to help prevent the advance of the agricultural frontier into indigenous peoples' forest homelands damaged by the disaster.
- Provide communities with more awareness-raising campaigns and training for disaster risk reduction; facilitate communities' access to resources that will help them adapt, including alternative production systems that can better cope with climate change and variability.
- Require national institutions (including local governments) to prepare and update emergency and disaster risk-reduction plans, to implement construction codes for disaster-prone areas, and to demarcate and title indigenous peoples' territories.
- Strengthen the preparedness of internal and external institutions that deal with indigenous communities by building their practical capacity to respond effectively to drastic climate changes and to help their communities face the impacts of these changes.

Needs for Further Research

During the preparation of this report, four broad topics emerged as areas in need of further research to study how indigenous peoples' vulnerability to poverty and migration is affected by climate change. How can their own institutions and knowledge systems be applied as a strategy to strengthen community resilience and support adaptation and mitigation initiatives? The key research areas emerging from this study include the following:

How climate change affects the vulnerability of indigenous peoples in LAC: poverty, assets, livelihood strategies, and the changing role of their cultural institutions. Preliminary empirical findings indicate that these cultural institutions are vital in shaping natural resource management and adaptation. Thus far, however, only limited knowledge is available on their specific role in the context of climate change.

How to increase the resilience of vulnerable indigenous communities to climate change by strengthening the capacity of their institutions. Measures for this purpose might include: (a) altering their production systems and encouraging alternative uses of natural resources to increase food security, (b) creating conflict-resolution mechanisms to resolve disputes over resources, and (c) identifying and advocating policy alternatives that enhance adaptive capacity.

How to incorporate indigenous people's knowledge into the design of instruments to mitigate and adapt to climate change. In particular, there is a need to address the role of indigenous peoples' informal institutions in community-level negotiations with government about (a) adaptation and mitigation initiatives, and (b) the use and conservation of natural resources. An important aspect will be to examine how to incorporate indigenous peoples' traditional knowledge into broader local as well as government-supported mitigation and adaptation schemes.

How to understand the migration drivers for indigenous peoples. Like other vulnerable groups, indigenous peoples are diverse and linked to access to economic, cultural, and social capital. Further study may reveal ways to increase the capacity of communities to cope with enhanced and new migration flows through a better understanding of both the beneficial effects of climate change (increased access to different knowledge resources, monetary means) and the negative effects (drain of youth from indigenous communities, reduced workforce with potentially decreased productivity, limited participation in traditional institutions, and weakened processes for transfer and development of indigenous peoples' knowledge).

Notes

1. See Orlove, Chiang, and Cane (2000) for further analyses on traditional climate knowledge in the Andes, and Kronik and Verner (2009) for a review of the role of indigenous peoples' knowledge in crafting adaptation strategies and mitigation instruments in LAC.

2. The World Bank is studying the forest sector and REDD. It would seem a natural follow-up to this volume to initiate a similar piece on the social impacts associated with climate-change mitigation.

References

Kronik, Jakob. 2010. *Living Knowledge—The Making of Knowledge about Biodiversity among Indigenous Peoples in the Colombian Amazon.* Saarbrücken, Germany: Lambert Academic Publishing.

Kronik, Jakob, and Dorte Verner. 2009. "The Role of Indigenous Knowledge in Crafting Adaptation and Mitigation Strategies to Climate Change in Latin America." In *Social Dimensions of Climate Change: Equity and Vulnerability in a Warming World*, eds. Robin Mearns, Andrew Norton, and Edward Cameron. Washington, DC: World Bank.

Orlove, B. S., J. C. H. Chiang, and M. A. Cane. 2000. "Forecasting Andean Rainfall and Crop Yield from the Influence of El Niño Pleiades Visibility." *Nature* 403: 68–71.

Perch-Nielsen, S., M. B. Bättig, and D. Imboden. 2008. "Exploring the Link between Climate Change and Migration." *Climatic Change* 91: 375–393.

Verner, Dorte. 2010. *Reducing Poverty, Protecting Livelihoods and Building Assets in a Changing Climate: Social Implications of Climate Change in Latin America and the Caribbean.* Washington, DC: World Bank.

Climate Change and Climatic Variability in Latin America and the Caribbean[1]

Warming of the climate system is unequivocal, as is now evident from observations of increases in global average air and ocean temperatures, widespread melting of snow and ice, and rising global average sea level. . . Continued greenhouse gas emissions at or above current rates would cause further warming and induce many changes in the global climate system during the 21st century that would very likely be larger than those observed during the 20th century. (IPCC 2007)

For LAC, information about long-term variations as well as more recent changes in climate is essentially absent from the international literature. The existing data series are traditionally safeguarded by national weather services and other regional authorities, which do not necessarily have routine-based procedures to make data easily accessible to external users. Only observations from a sparsely distributed network of stations are available for in-depth analysis. The paucity of information has so far limited not only studies of observed climate change and variability, but also projections of climate change and variability in the region. Thus, the quality of climate models pertaining to LAC is difficult to assess.

This appendix summarizes knowledge on recent climate change and variability and the most-likely future climate change and variability in

LAC, building primarily on the 2007 Intergovernmental Panel on Climate Change (IPCC) Fourth Assessment reports and recent literature.

LAC is warming largely in line with the global trend and is likely to continue doing so. There will be local exceptions to the general tendency, but available models do not give robust results for every part of the region. In particular, there are still many unresolved issues related to changes in the Amazonas, because important aspects of the interaction between vegetation and climate are still little understood. Different models also tend to behave differently in simulating the present climate within the region—which limits our ability to take the simulated responses to anthropogenic forcings at their face value for the region. For this reason, the best estimate of climate change and variability within the region comes from assessing the results from many models, as was done by Christensen and others (IPCC 2007). The present analysis tries to go behind some of the statements provided there by undertaking a comparative assessment of the results of general circulation models (GCMs).

Images of Present Change

Temperature
As noted above, LAC is projected to continue to warm at a rate little different from that of the world as a whole. Figure A.1 shows the geographical distribution of linear temperature trends for the periods 1901–2005 and 1979–2005. The temperature trend for 1901–2005 shows that larger geographical variations and noticeable "hot spots" are in southeast Brazil, Uruguay, northeast Argentina, and northwest Mexico, where the warming has been more than double the global increase (box A.1). For the more recent period, the warming trend shows less geographical variance.

Precipitation
Data are less widely available for precipitation than for temperature, but the available data show that precipitation varies widely within the LAC region. Figure A.2 depicts the geographical distribution of change. The most noticeable large-scale coherent pattern is that of a long-term tendency toward drying in the tropics and subtropics (the Caribbean and northern South America), while temperate climates (southern South America) experience more precipitation.

Figure A.1   Linear Trend of Annual Temperatures

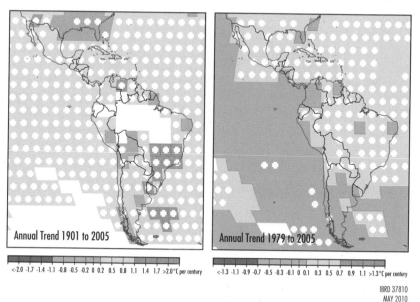

Source: Adapted from IPCC 2007.
Note: Areas in grey have insufficient data to produce reliable trends. Trends significant at the 5 percent level are indicated by white + marks.

For large parts of the region, data are not readily available for analysis. In some countries, efforts have been made to rescue and collect observational data, though not necessarily to make them publicly available. Analysis of these data confirms the broad tendencies, but trends may vary quite widely even at the community level (scale of 1–200 km), with quite large positive trends occurring next to negative ones. It is only when data are aggregated over large regions that changes can be described in the context of global warming (Zhang and others 2007).

Sea Level

Sea level in many regions of the world varies considerably on many time scales. The shortest time variations and in some regions the largest sea-level changes by far are induced by tides. However, as these occur on a regular basis, natural and managed environments are clearly adapted to this variability.

Globally, sea level has been rising over the centuries, and during the 20th century an average increase of 0.17 meters was measured (IPCC 2007).

Box A.1

Climate Definitions

Often there is a considerable gap between common perceptions of "climate" and a more scientific, meteorologically based definition. This mismatch often gives rise to misleading statements and assessments of the role of mankind as a driver of observed changes, even in high-level documents. To better guide the reader of the present volume, we provide here a basic meteorological understanding of what climate is.

Annual Global Mean Temperatures

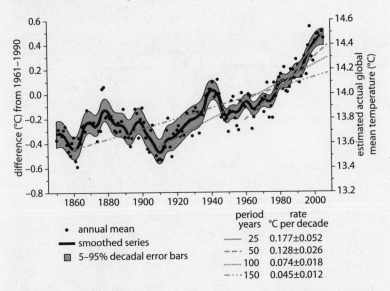

	period years	rate °C per decade
• annual mean		
━━ smoothed series	—— 25	0.177±0.052
■ 5–95% decadal error bars	—·— 50	0.128±0.026
	········ 100	0.074±0.018
	—··— 150	0.045±0.012

Source: Adapted from IPCC 2007.
Note: Annual global mean temperature (black dots) with linear fits to the data. The left-hand axis shows temperature anomalies relative to the 1961–1990 normal period, and the right-hand axis shows estimated actual temperatures, both in °C. Linear trends are shown for the last 25 (solid), 50 (dashed), 100 (dotted), and 150 (dashes and dots) years. The smooth blue curve shows decadal variations, with the decadal 90% error range shown as a pale blue band about that line. The total temperature increase from the period 1850–1899 to the period 2001–2005 is 0.76°C ± 0.19°C.

Climate is basically the statistical properties of weather over a long period. The simplest and best-understood parameters examined are mean temperature and

(continued)

Box A.1 *(continued)*

mean precipitation amounts, whether on a monthly, seasonal, or annual basis. No law of physics can tell exactly which time frame to use. When weather changes from one year to another and possibly shows significant trends over some years, it is not in agreement with the above definition of climate. A long time period must be defined so that comparisons between two climate periods (two equally long periods) are close to invariant. This means that year-to-year or even decadal variations should show up only marginally. However, even at longer time periods, the statistics of weather do not seem to be entirely invariant. In practice, climatologists in the first part of the 20th century decided to use a period of 30 years as a compromise to balance the need for invariance in the conditions from one period to another. This led to the definition of 30-year climate norms, which started with the period 1901–1930. The latest norm is for the period 1961–1990. For practical reasons, these periods are still used with rigor. Therefore, many climate variables such as annual temperature and precipitation are compared with respect to the latest reference period. As climate turns out to vary significantly even between these so-called norm periods, it also has become customary to look at long-term trends.

The figure above shows a time series of annual global mean temperatures from 1850 to 2005. The overall increasing trend is portrayed in a number of ways (see figure annotation). It is important to note that the underlying data points show a clear scatter around the overall trend. Trying to interpret the time evolution over too short a period is made difficult and misleading by the considerable interannual "noise." For example, trying to make a linear trend line through the last 8–10 years (that is, from 1998 to the present) would result in a close to zero or even negative trend, suggesting that the overall warming has stagnated. Such interim periods have occurred previously. It is only with a longer time span that the real picture of change clearly shows itself. Change need not be a constantly evolving phenomenon; even a reverse signal at times need not contradict the long-term evolution.

The definition of climate change and variability is not just a question of comparing climate periods; in general it must also involve regions large enough to show a coherent picture of change. Only then is it possible to relate the climate trend of the site to any large-scale change taking place in the region of concern. The geographical scale on which such coherence is found appears to be not much finer than continental. Therefore, clear-cut and robust statements about observed climate change and variability on a detailed country basis would be dubious, if they are not balanced with a view to the larger-scale trends both geographically and temporally. Opposite trends at nearby stations may still be in accordance with changes on a larger scale.

Source: Author.

Figure A.2   Trend in Annual Land Precipitation Amounts, 1901–2005

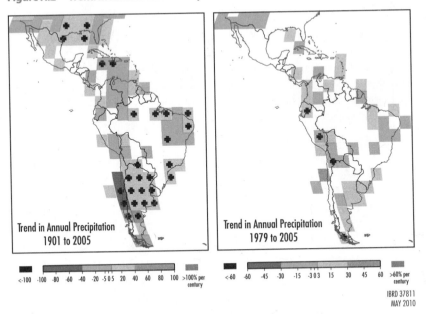

IBRD 37811
MAY 2010

Source: Adapted from IPCC 2007.
Note: Areas in grey have insufficient data to produce reliable trends. Note the different color bars and units in each plot. Trends significant at the 5 percent level are indicated by black + marks.

Regionally, however, there are considerable variations (figure A.3). Oceanic circulation is complex, and there is little reason to believe that the observed geographical distribution of these changes can be directly scaled to climate change and variability scenarios, because the changing ocean conditions (that is, currents, salinity, and temperature) may result in less obvious results along the way. However, past experience often serves as a reference point against which to compare projections of the future.

Recent Events

Most of LAC has experienced several instances of severe weather in recent decades, damaging property, infrastructure, and natural resources. Fatalities in connection with torrential rain and hurricanes have been counted in thousands within this century alone. The overall perception of these events is that they signal climate change and variability. However, there is no formal scientific detection of these changes at the regional level. This makes it difficult to assess whether these events—either individually or collectively—are a result of general climate change and variability indicated by the overall warming of the continent and adjacent areas.

Figure A.3   Geographic Distribution of Long-Term Linear Trends in Mean Sea Level, 1955–2003

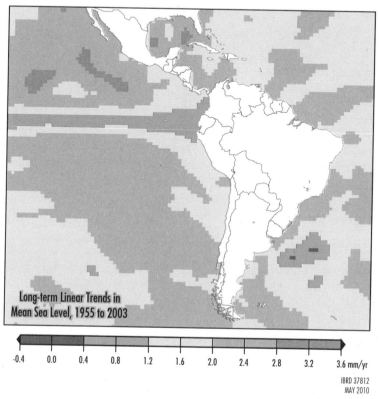

Long-term Linear Trends in Mean Sea Level, 1955 to 2003

-0.4   0.0   0.4   0.8   1.2   1.6   2.0   2.4   2.8   3.2   3.6 mm/yr

IBRD 37812
MAY 2010

Source: Adapted from IPCC 2007.
Note: Based on the past sea-level reconstruction with tide gauges and altimetry data.

People who have experienced severe and damaging weather events generally wish to assign the events to some special factor. Yet changes in climate or the recurrence of extreme events need not be related to global warming. To formally attribute change or the occurrence of particular events to a cause, a statistically sound number of events normally must be considered. By nature, extreme events are rare at any given location. Even events that occur over a large geographical region such as LAC cannot simply be aggregated and studied as a whole, because the chain of physical events leading to any one of these is likely to differ from event to event and from location to location. This precludes a simple statistical treatment of the data. Therefore, recent IPCC reports have had very little to say about recent changes at a regional level—and even less at the national or provincial level—and their possible links with global climate change and variability.

Nonetheless, people relate the impacts of new events to the memory of past events. And as adverse changes create a need for adaptation and mitigation measures, it is important to use past experience in preparing for potential hazards.

Images of the Future

Global climate models have long been the main tool used to project climate change and variability under certain assumptions about future emissions of greenhouse gases (GHGs) and other anthropogenically induced drivers of the climate system. The most commonly used climate variable to illustrate anthropogenically induced climate change is the global annual mean temperature. Observations show that the average global temperature rose by 0.74°C over the last century (IPCC 2007). Model experiments show that even if all human-induced forcing agents were held constant at their year 2000 levels, further warming would occur in the next two decades at a rate of about 0.1°C per decade, mainly reflecting the slow response of the oceans. About twice as much warming (0.2°C per decade) would be expected if emissions are within the range of the IPCC's *Special Report on Emissions Scenarios* (SRES) (Nakicenovic and others 2000). Best-estimate projections from models indicate that the average warming over each inhabited continent by 2030 is insensitive to the choice among scenarios. Moreover, according to IPCC (2007), average warming is very likely to be at least twice as large as the corresponding model-estimated natural variability that took place during the 20th century.

Best estimates and likely ranges for global average surface air warming for six SRES emissions marker scenarios were provided by IPCC (2007). For example, the best estimate for the low scenario (B1) is 1.8°C (likely range is 1.1°C to 2.9°C), and the best estimate for the high scenario (A1FI) is 4.0°C (likely range is 2.4°C to 6.4°C).[2]

The SRES estimates refer to the entire planet and hide large regional variations. The spread between different model results that is indicated by the quoted range of the two marker scenarios B1 and A1FI adds to this uncertainty. Although different models tend to respond to an anthropogenic forcing largely in the same manner on the global scale, the spread is a clear signal of large discrepancies on the regional scale over many regions.

Climate varies from region to region. This variation is driven by the uneven distribution of solar heating; available atmospheric moisture; the individual responses of the atmosphere, oceans, and land surface; the interactions among these; and the physical characteristics of the regions. The perturbations of the atmospheric constituents that lead to global changes affect certain aspects of these complex interactions. Some anthropogenic forcings are global in nature, while others differ regionally. For example, carbon dioxide, which causes warming, is distributed evenly around the globe regardless of where the emissions originate, whereas sulfate aerosols (small particles), which offset some of the warming, tend to be regional in their distribution. Furthermore, the response to forcings is partly governed by feedback processes that may operate in regions different from those in which the forcing is greatest. Thus, the projected changes in climate will also vary from region to region.

Against this backdrop, the array of models that yield robust projections of the global rise in mean temperature yield much less robust estimates of *regional* climate change and variability (IPCC 2007). To the extent assessments of future climate change and variability can be made, they are as follows:

- All of Central and South America is very likely to warm during this century. The annual mean warming in southern South America is likely to be similar to the global mean warming, but larger than the mean warming in the rest of the area.

- Annual precipitation is likely to decrease in most of Central America, with the relatively dry boreal spring becoming drier. Annual precipitation is likely to decrease in the southern Andes, with relative precipitation changes being largest in summer. A caveat at the local scale is that changes in atmospheric circulation may induce large local variability in precipitation changes in mountainous areas. Precipitation is likely to increase in Tierra del Fuego during winter and in southeastern South America during summer.

- It is uncertain how annual and seasonal mean rainfall will change over northern South America, including the Amazon forest. In some regions, there is qualitative consistency among the simulations: rainfall increasing in Ecuador and northern Peru, and decreasing at the northern tip of the continent and in the southern portion of northeast Brazil.

- The systematic errors in simulating current mean tropical climate and its variability, and the large intermodel differences in future changes in El Niño amplitude, preclude a conclusive assessment of the regional changes over large areas of Central and South America. Most models are poor at reproducing the regional precipitation patterns in their control experiments and have a small signal-to-noise ratio, in particular over most of Amazonia. The high and sharp Andes Mountains are unresolved in low-resolution models, affecting the assessment over much of the continent. As with all landmasses, the feedbacks from land-use and land-cover change are not well accommodated and lend some degree of uncertainty. The potential for abrupt changes in biogeochemical systems in Amazonia remains a source of uncertainty. Large differences in the projected climate sensitivities in the climate models incorporating these processes and a lack of understanding of processes have been identified. Over Central America, tropical cyclones may become an additional source of uncertainty for regional scenarios of climate change and variability, because the summer precipitation over this region may be affected by systematic changes in hurricane tracks and intensity.

The following section interprets these findings on a regional scale.

Aspects of Observed Climate

Ideally, GCMs should be able to provide information at the regional scale on which they resolve, but efforts to improve such models have concentrated on the ability to describe specific geophysical phenomena such as El Niño, monsoon systems, and sea ice. This focus has precluded paying specific attention to certain aspects of model performance on a regional level for many parts of the world, including LAC.

Therefore, alternative methods have been developed to derive detailed regional information in response to geophysical processes at finer scales than resolved by GCMs. Nested regional climate models (RCMs) and empirical downscaling have yielded new ways to assess important regional processes that are central to climate change and variability. These assessments allow the development and validation of models to simulate the key dynamic and physical processes of the climate system. Until recently, there were few scientific programs aiming at providing high-resolution information on climate change in developing nations, including those of LAC. One exception is the Japanese initiative, the Earth Simulator. Here, information from a dedicated simulation with a high-resolution GCM is providing fine-scale information globally, including for LAC. However, results from this simulation should be carefully compared with the collective information from other GCMs (IPCC 2007) as information from a single simulation will provide only indicative information about possible future change.

Within the community studying climate change impacts and adaptation there is a growing move toward integrated assessment, yielding projections of regional climate change and variability that provide key inputs into decision-support systems aimed at reducing vulnerability (Bales, Liverman, and Morehouse 2004). At present, the regional projections are perhaps the weakest link in this integrated assessment, and the bulk of information readily available for policy and resource managers (such as via the IPCC Data Distribution Center[3]) largely derives from GCMs, which have limited ability to accurately simulate local-scale climates, especially for the key parameter of precipitation. GCM data are commonly mapped as continuous fields, which do not convey the low skill of the models for many regions, or are area aggregated, which renders the results of little value for local application.

To help understand the potential accuracy of climate change and variability projections derived from climate models, it is imperative to compare the performance of these models against the observed climate. Below we summarize some of the main LAC climate characteristics, which a model should be able to describe with some realism if its projections for the future are to be credible.

Mexico and Central America
Most of the American isthmus (for example, central-southern Mexico and Central America) and the Caribbean have a relatively dry winter and a well-defined rainy season from May through October (Magaña, Amador, and Medina 1999; Taylor and Alfaro 2005). The progression of the rainy season largely results from air–sea interactions over the Americas' warm pools (for example, the Gulf of Mexico) and the effects of topography over a dominant easterly flow, as well as the temporal evolution of the Inter-Tropical Convergence Zone (ITCZ).[4] The mountain range running the length of the American isthmus limits moist air moving across it, resulting in a precipitation pattern characteristic of mountain ranges. As moist air that has formed over the sea moves in over land, it is forced up the mountainside and, as it cools down, releases rainfall on the ascent. The high mountains and the lee-side of the mountains remain largely dry. This effect is present whether the air flows from east to west or vice versa. During the boreal winter, the atmospheric circulation over the Gulf of Mexico and the Caribbean Sea—that is, the Intra Americas Seas (IAS)—is dominated by

the seasonal fluctuation of the Subtropical North Atlantic Anticyclone, with invasions of extratropical systems that affect mainly Mexico and the western portion of the Great Antilles. The so-called *Nortes* or *Tehuantepecers* produce some precipitation and changes in temperature over the coastal regions of the IAS (Romero-Centeno and others 2003).

The rainy season in the American isthmus and the Caribbean is mainly the result of easterly waves and tropical cyclones that contribute to a large percentage of the precipitation. When low vertical wind shear coincides with warm sea-surface temperatures, two of the necessary preconditions exist that may result in easterly waves maturing into storms and hurricanes, generally within the 10°N–20°N latitudinal band.

Over most of the IAS, warm pool precipitation is weak because subsidence (descending air masses) inhibits convective activity, making this region a climate paradox. Over the Caribbean Sea, winds are strong because of a low-level jet, which seems to play a key role in the distribution of precipitation over Central America. A unique characteristic of boreal summer precipitation over the Pacific side of the American isthmus and the adjacent warm pool is its bimodal structure, with maxima in June and September and a relative minimum during the middle of the boreal summer, that is, late July and early August. This relative minimum, known as *Canicula* or Midsummer Drought, has been attributed to air–sea interactions and teleconnections between the IAS and the eastern Pacific warm pool (Magaña, Amador, and Medina 1999). This characteristic in convective activity on these times scales appears to influence tropical cyclone formation in the eastern Pacific.

Most climate variability in the region is related to the El Niño Southern Oscillation (ENSO), a phenomenon that influences the distribution, frequency, and intensity of many of the regional atmospheric phenomena. The signal of El Niño over the American isthmus and the Caribbean is contrasting, not only among seasons but also in relation to coast (Pacific or Caribbean). For instance, during El Niño boreal winters, precipitation increases over northwestern Mexico, the Greater Antilles, and part of the Pacific Coast of Central America, and less rainfall than usual is observed in parts of Colombia and Venezuela and the Lesser Antilles. During the El Niño boreal summers, most of the American isthmus experiences negative precipitation anomalies, except along the Caribbean Coast of Central America, where positive precipitation anomalies are observed. Tropical cyclone activity over the IAS diminishes during El Niño summers (Gray 1984; Tang and Neelin 2004).

The mean state of ENSO and its global pattern of influence, amplitude, and interannual variability and frequency of extreme events have varied considerably in the past. Many of these changes appear to be related to, though definitely not entirely to the result of, changes in global climate and the history of external forcing agents, including recent anthropogenic forcing (Mann, Bradley, and Hughes 2000). Interannual and decadal ENSO-like climate variations in the Pacific Ocean basin are important contributors to the year-to-year (and longer) variations of the climate in South (and North) America. Despite potentially different source mechanisms, both interannual and decadal ENSO-like climate variations yield wetter subtropics (when the ENSO-like indices are in positive, El Niño–like phases) and drier midlatitudes and tropics (overall) over the Americas, in response to equator-ward shifts in westerly winds and storm tracks in both hemispheres (Dettinger and others 2001).

South America

A complex variety of regional and remote factors contributes to the climate of South America (Nogues-Paegle and others 2002). The tropospheric upper levels are characterized by high pressure centered over Bolivia and low pressure over northeast Brazil. At low levels, the Andes effectively block air exchanges with the Pacific Ocean, but a continental-scale gyre transports moisture from the tropical Atlantic Ocean to the Amazon region, and then southward toward extratropical South America. The South American low-level jet starts a regional intensification of this flow, channeling it along the eastern foothills of the Andes into the so-called Chaco low.[5] The low-level jet carries significant moisture from the Amazonas toward southern South America and is present throughout the year, but strongest during the austral winter season (Berbery and Collini 2000; Marengo and others 2004; Vernekar, Kirtman, and Fennessy 2003).

A clear warm season precipitation maximum, associated with the South American Monsoon System (SAMS), dominates the mean seasonal cycle of precipitation in tropical and subtropical latitudes. The rainfall over northern South America is directly influenced by east–west circulation patterns, and consequently tropical sea-surface temperature anomalies affect regions such as the Ecuador coast and north-northeast Brazil.[6] The SAMS is also modulated by incursions of drier and cooler air from the midlatitudes over the interior of subtropical South America (Garreaud 2000; Vera and Vigliarolo 2000). Rainfall anomalies over subtropical South America are associated with regional feedback processes and interactions among the topography, the SAMS, and the midlatitude systems.

Another important feature, a regional part of the ITCZ, is the South Atlantic Convergence Zone (SACZ)—a southeastward extension of cloudiness and precipitation from southern Amazonas toward southeast

Brazil and the neighboring Atlantic Ocean. The SACZ reaches its easternmost position during December, in association with high precipitation over much of Brazil, a southeasterly flow over eastern Bolivia, and low precipitation in the Altiplano. The variability of precipitation during the austral summer on a variety of time scales (intraseasonal, interannual) has been related to changes in the position and intensity of the SACZ (Liebmann and others 1999). Variability of the SACZ also influences the atmospheric circulation and rainfall anomalies over eastern South America, between about 20°S and 40°S, through a dipole pattern in the vertical motion field that reflects changes in the intensity of the SACZ and "compensating" descent over southern Brazil, Uruguay, and northeastern Argentina (Doyle and Barros 2002; Robertson and Mechoso 2000).

The leading mode of the interannual variability in the Southern Hemisphere is the Southern Annual Mode (SAM) (Kidson 1988; Thompson and Wallace 2000). For southeastern South America, results indicate that the SAM activity modulates the regional variability of the precipitation, especially during the late austral spring, when the SAM index correlates well with ENSO (Silvestri and Vera 2003).

Global Climate Change Issues

The scientific attribution of climate change and variability to causes calls for a comprehensive understanding of natural variability and large-scale external drivers such as GHG concentrations and aerosol loads. By combining information on observed changes with information from specially designed experiments using climate models with and without the external drivers, it becomes possible to distinguish a possible effect originating from the driver and, hence, to attribute the cause of an observed change to the driver. However, just as models cannot perfectly represent real climate, observations cannot precisely portray climate evolution. For this reason, it is very difficult to precisely define what models should depict and what should be understood as merely an artifact of natural variability, reflecting processes that are basically unaffected by an external driver.

Climate variability exists on many time scales and on many geographical scales. To illustrate the difficulty of detecting and attributing climate change and variability, the occurrence of El Niño provides a good example. El Niño occurs at approximately four-year intervals. The physical chain of reactions leading to its occurrence is well understood (IPCC 2007), and state-of-the-art climate models are generally able to simulate both El Niño and La Niña episodes with fidelity. But to simulate the observed sequence of events, models need to know the precise ocean state at certain intervals. Given the chaotic behavior of the climate system, the ocean component of a climate model will drift away from the exact development within a few years of simulations, even if it is initialized with the best observed state of the ocean. The same behavior is seen in the atmospheric component because categorical weather forecasts beyond two to three weeks are impossible.

Since it is impossible to capture the precise evolution of important regional climate features such as El Niño and La Niña, models will not generally be able to describe the transient trends and variability associated with such large-scale natural drivers. This should not be seen as a limitation, but acknowledging such general facts provides some insight on how to assess the results from climate change models.

For models, the depiction of regional features of change becomes increasingly demanding when one wants to study the smaller scale. This is a combination of the limited resolution presently used in most models and the nature of natural climate variability, which may not be as well characterized as El Niño or La Niña either in temporal regularity or in extent. It is therefore interesting that, collectively, models can simulate the temperature trends of the 20th century not only on the global but also on a continental scale. Figure A.4 makes this clear and also shows that the rising trends can be interpreted only as the result of anthropogenic drivers of the climate system.

Figure A.4 also shows that the models do not capture very well the ocean surface temperatures during a short period in the early part of the 20th century. As a result, even though the models are better at simulating land surface temperatures for that same period, the observed increase in global surface temperatures for that period are less well captured. Therefore, this discrepancy should be looked for in models' ability to simulate the accurate ocean surface temperatures. Because detailed ocean temperatures are the result of a complex dependency on the centuries-old oceanic circulation and the immediate climate forcing, it is hardly a surprise that there is a mismatch at a time when the external climate forcing was small.

To what extent can or should models be expected to simulate the climate evolution in LAC? Given the arguments above, there is no reason to expect models to be able to reproduce natural variations in detail, with the exception of the large-scale temperature increase. It is therefore important to keep in mind that only if the externally induced climate

Figure A.4 Comparison of Observed Continental- and Global-Scale Changes in Surface Temperature

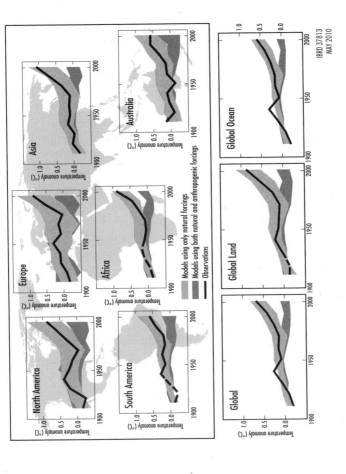

Source: Adapted from IPCC 2007.

Note: Results simulated by climate models using natural and anthropogenic forcings. Decadal averages of observations are shown for 1906 to 2005 (black line) plotted against the center of the decade and relative to the corresponding average for 1901 to 1950. Lines are dashed where spatial coverage is less than 50 percent. Gray bands show the 5 percent to 95 percent range for 19 simulations from five climate models using only the natural forcings from solar activity and volcanoes. Green bands show the 5 percent to 95 percent range for 58 simulations from 14 climate models using both natural and anthropogenic forcings.

change and variability is sufficiently large will models depict a signal above "climate noise" or natural variability. For this reason alone, most efforts to model regional climate change and variability focus on scenarios where a large climate response should be expected. Moreover, they focus on the century-long time scale and not on the near term, even though the near-term horizon is where the need to know the evolution may seem most urgent.

The implication of these findings for interpreting locally observed recent changes in temperature and other climate variables is that they are subject to doubt. Observations may even suggest that local change is opposite to the global- or continental-scale behavior. This does not disprove global or regional climate change and variability, but it does remind us that even if global temperatures should increase by 3–4°C within this century, many sites may not experience any of this change, while others will see an even greater change.

Projections of Regional Climate Change

Region-by-region projections of climate change and variability need to be interpreted based on a proper understanding of key regional processes and of the skill of models in simulating current regional climate (box A.2).

It is imperative for the use of the information provided at the regional level, and even more so at a country level, to see the statements offered in this appendix against the background of uncertainties related to the issue of regional climate change and variability. Largely because observational records do not have a long and uninterrupted time series (going back to the beginning of the 20th century, for example), there is no evidence that what we have seen will be less or more severe in comparison with what we may experience within this century. Observed climate trends and the occurrence of extreme events in LAC cannot generally be used to deduce information about things to come. On the other hand, models clearly suggest that changes will take place over this century that will generally be in the direction of more of the extremes, that is, more intensive precipitation, longer dry spells, warm spells, and heat waves with higher temperatures than generally experienced up to now.

It has often been stated that observed changes in adverse climate events already show characteristics that models predict for the future. However, this cannot be statistically confirmed because the intrinsic rarity of extreme events prevents them from being properly sampled. The conclusion drawn must be that we may not have seen the worst yet, and

Box A.2

Sources of Uncertainty in Regional Climate Change Projections

Climate models are constructed to match observed climatic conditions, which they do with varying degrees of accuracy. When applied in century-long climate change and variability projections, these differences result in deviating responses to the prescribed climate forcing resulting from specified levels of GHGs. This introduces uncertainty in the climate projections.

We do not know with any certainty the future emissions of GHGs; this introduces uncertainty in climate change and variability projections on time scales beyond a few decades.

Natural variability in present climate conditions introduces a certain background noise, which any projection of climate change and variability has to be compared with. Depending on the amplitude of this variability, the actual change may be difficult to filter out, as the "signal" remains below the noise. This is particularly relevant on the regional and local scales because the amplitude of variability generally increases as the scale gets smaller.

Weather phenomena that are climate specific but appear only irregularly or very rarely—for example, extreme events—happen so rarely that a formal analysis of change is impossible because of the lack of a proper statistical basis for analysis. This is particularly relevant on the regional and finer scales.

While climate change and variability projections for the distant future are independent of the precise state of current climate, this is not the case for projections covering the next one to two decades. Currently, imperfect knowledge, primarily about the state of the oceans, precludes prediction of seasonal weather patterns much beyond six months, even on the very large scales.

Climate change and variability projections for the near future—despite being independent of the emission scenario—generally become very uncertain due to the signal-to-noise issue described above and the limited knowledge we possess about the exact current state of the climate.

Robust statements about change on the regional scale are therefore possible only if the model projections also are physically sound, meaning that a certain effect is expected because of large-scale changes in atmospheric circulation, moisture content, or temperature change. These measures of change should be captured by most if not all climate models. In most cases, it is therefore not possible to provide formal quantitative estimates of error for the projected values of change.

Source: Author.

that preparations to deal with existing climatic hazards are needed as a first step toward coping with things to come. This statement is generic and particularly applicable to LAC, given the so-far limited scientifically scrutinized analysis.

The following discussion is organized according to the regions used in IPCC (2007), largely covering Central and South America and some of the islands in the Caribbean. These regions are continental scale (or based on large oceanic regions with a high density of inhabited islands); they may have a broad range of climates and be affected by a large range of climate processes. As they are generally too large to be used as a basis for conveying quantitative information on regional climate change and variability, the LAC region is further divided into Southern South America (SSA), Amazon (AMX), and Central America and Mexico (CAM).

This regionalization is very close to that initially devised by Giorgi and Francesco (2000), but it includes additional oceanic regions and some other minor modifications similar to those of Ruosteenoja and others (2003). The objectives behind the original Giorgi and Francesco regions were that they should have simple shape, be no smaller than the horizontal wave length typically resolved by GCMs (judged to be a few thousand kilometers), and, where possible, should recognize distinct climatic regimes. Although these objectives may be met with alternative regional configurations, as yet there are no well-developed options in the regional climate change and variability literature.

Several common processes underlie climate change and variability in a number of regions. Before discussing LAC countries individually, it is relevant to summarize some of the common processes. The first is a fundamental consequence of warmer temperatures and the increase in water vapor in the atmosphere. Water is continually transported horizontally by the atmosphere from regions of moisture divergence (particularly in the subtropics) to regions of convergence. Even if the circulation does not change, these transports will increase because of the increase in vapor, and regions of convergence will get wetter and regions of divergence drier. We see the consequences of this increased moisture transport in plots of the global response of precipitation—where, on average, precipitation increases in the intertropical convergence zones, decreases in the subtropics, and increases in subpolar and polar regions (Meehl and others 2007). Regions of large uncertainty often lie near the boundaries between robust moistening and drying regions, with different models placing these boundaries differently, as illustrated by figure A.5.

Figure A.5   Temperature and Precipitation Changes over Central and South America

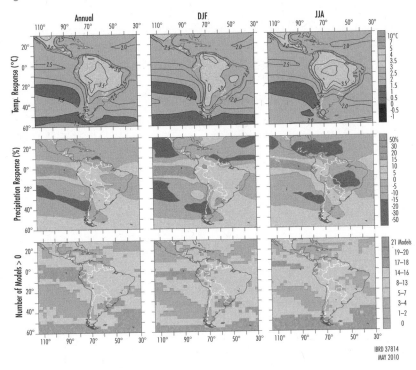

Source: Adapted from IPCC 2007.
Note: Top row: Annual mean, December-January-February and June-July-August temperature change between 1980–99 and 2080–99, averaged over 21 models. Middle row: Same as top, but for percentage change in precipitation. Bottom row: Number of models out of 21 that project increases in precipitation.

Two other important themes in the extratropics are the poleward expansion of the subtropical high-pressure systems responsible for the overall long-term stable and dry weather conditions, and the poleward displacement of the midlatitude westerly wind bands with the associated storm tracks. This atmospheric circulation response is often referred to as the excitation of the positive phase of the SAM. Superposition of the tendency toward subtropical drying and poleward expansion of the subtropical highs create especially robust drying responses on the equatorward boundaries of the subtropical oceanic high centers.

The generally poor ability of climate models to simulate present climate for the region, incomplete analyses of climate change and variability in global climate model results for LAC, and a serious lack of detailed studies of downscaled climate information enable relatively few robust statements to be made (IPCC 2007a).

Nonetheless, drawing from the analysis of many climate models (IPPC 2007; Meehl and others 2007), some clear indications about regional changes do appear to be rather robust. Figure A.5 summarizes the collective information about temperature and precipitation changes toward the end of the current century compared with the current climate, based on 21 GCMs used by IPCC (2007).

Temperature

The balance of evidence assessed by the IPCC (2007) leads to the following statement:

- All of Central and South America is very likely to warm during this century. The annual mean warming is likely to be similar to the global mean warming in southern South America but larger than the global mean warming in the rest of the area.

This implies that temperatures in all seasons will continue to rise during the 21st century. Some aspects of temperature-related events are also expected to change. Unless the temperature increase is a result of entirely new circulation patterns, changes in heat wave frequency and intensity are expected. Likewise, there is an enhanced risk of changes in the seasonality of some severe events because a higher temperature level in general will favor a longer warm season with possible related extreme events. A clear example is that of the hurricane season (box A.3).

Box A.3

Hurricanes

An important driver for tropical cyclones and hence hurricanes is the sea surface temperature (SST). A threshold at about 26°C determines whether a hurricane can be formed. In a warmer world with everything else being equal, the hurricane season is likely to be prolonged, and the area prone to hurricane development may expand. However, there will always remain a zone free of such systems near the equator, where atmospheric motions cannot support the development of intensive storms. March 2004 saw the first hurricane ever identified in the southern Atlantic. This hurricane severely affected the coastal zone of northeast Brazil. Such systems could become more frequent as a result of higher SST in the region.

In the tropics, particularly when the sun is close to its zenith, outgoing long-wave radiative cooling from the surface to space is not effective in the optically

(continued)

Box A.3 *(continued)*

thick environment caused by the high water vapors over the oceans. Links to higher latitudes are weakest at this time, and transport of energy by the atmosphere, such as occurs when the sun is less strong, is not an effective cooling mechanism, while monsoonal circulations between land and ocean redistribute energy in areas where they are active. However, tropical storms cool the ocean surface through mixing with cooler, deeper ocean layers and through evaporation. When the latent heat is realized in precipitation in the storms, the energy is transported high into the troposphere where it can radiate to space.

As the climate changes and SSTs continue to rise, the environment in which tropical storms form is changed. Higher SSTs are generally accompanied by increased water vapor in the lower troposphere, and thus the moist static energy that fuels convection and thunderstorms is also increased. As noted above, hurricanes generally form from preexisting disturbances only where SSTs exceed about 26°C; a rise in SST therefore potentially expands the areas over which such storms can form. However, many other environmental factors, including wind shear in the atmosphere, also influence the generation and tracks of disturbances. El Niño and variations in monsoons as well as other factors also affect where storms form and track. The potential intensity, defined as the maximum wind speed achievable in a given thermodynamic environment, similarly depends critically on SSTs and atmospheric structure (Emanuel 2003). Many factors, in addition to SSTs, determine whether convective complexes become organized as rotating storms and develop into full-blown hurricanes.

Although attention has often been focused simply on the frequency or number of storms, the intensity and duration likely matter more. The energy in a storm is proportional to velocity squared, and the power dissipation of a storm is proportional to the wind speed cubed. Consequently, the effects of these storms are highly nonlinear, and one big storm may have much greater impact on the environment and climate system than several smaller storms.

From an observational perspective, then, key issues are the tropical storm formation regions; the frequency, intensity, duration, and tracks of tropical storms; and the associated precipitation. For landfalling storms, the damage from winds and flooding, as well as storm surges, is especially of concern. But damage often depends more on human factors, including whether people place themselves in harm's way, their vulnerability, and their resilience achieved through such measures as building codes.

In summary, some models project hurricane intensities to increase, but there is little to document how their frequency and number of landfalls may develop.

Source: Author.

Precipitation

IPCC (2007) considered only a few features of future precipitation change in LAC to be robust. It stated the following:

- Annual precipitation is likely to decrease in most of Central America, with the relatively dry boreal spring becoming drier. Annual precipitation is likely to decrease in the southern Andes, with relative precipitation changes being largest in summer.

- Area mean precipitation in Central America decreases in most models in all seasons. It is only in some parts of northeastern Mexico and over the eastern Pacific during June, July, and August that increases in summer precipitation are projected. However, since tropical storms can contribute a significant fraction of the rainfall in the hurricane season in this region, these conclusions might be modified by the possibility of increased rainfall in storms not well captured by these global models.

- The annual mean precipitation is projected to decrease over northern South America near the Caribbean coasts, as well as over large parts of northern Brazil, Chile, and Patagonia, while it is projected to increase in Colombia, Ecuador, and Peru, around the equator, and in southeastern South America. The seasonal cycle modulates this mean change especially over the Amazon basin where monsoon precipitation increases in December, January, and February and decreases in June, July, and August. In other regions (for example, Pacific coasts of northern South America, and a region centered over Uruguay and Patagonia) the sign of the response is preserved throughout the seasonal cycle.

As shown in the bottom panels in figure A.5, models foresee a wetter climate near the Rio de la Plata and drier conditions along much of the southern Andes, especially in December, January, and February. However, when estimating the likelihood of this response, the qualitative consensus within this set of models must be weighed against the fact that most models cannot reproduce the regional precipitation patterns in their control experiments with much accuracy.

The poleward shift of the South Pacific and South Atlantic subtropical anticyclones is a very marked response across climate models. Parts of Chile and Patagonia are influenced by the polar boundary of the subtropical anticyclone in the South Pacific, and they experience particularly strong drying because of the combination of the poleward shift of circulation and increase in moisture divergence. The strength and position of the subtropical anticyclone in the South Atlantic is known to influence the climate of southeastern South America. The projected increase in rainfall in southeastern South America is related to a corresponding poleward shift of the Atlantic storm track.

Some projected changes in precipitation (such as the drying over east-central Amazonia and northeast Brazil and the wetter conditions over southeastern South America) could be a partial consequence of the El Niño–like response that is projected by many models (Meehl and others 2007).[7] This would directly affect tropical South America and affect southern South America through extratropical teleconnections (Mo and Nogués-Paegle 2001).

Extreme Events

Little research is available on extremes of temperature and precipitation for the LAC region. Christensen and others (IPCC 2007) estimate how frequently the seasonal temperature and precipitation extremes as simulated in 1980–99 are exceeded, using the A1B scenario[8] from a large ensemble of GCM simulations. They find that essentially all seasons and regions will be extremely warm by the end of the century. In Central America, the projected decrease in precipitation is accompanied by more frequent dry extremes in all seasons. In Amazonia, models project extremely wet seasons in about 27 percent of all summers (December, January, February) and 18 percent of all autumns (March, April, May) in the period 2080–99, with no significant change for the rest of the year. For southern South America, significant changes are not projected in the frequency of extremely wet or dry seasons. However, more careful analysis is required to determine how often these wet and dry extremes are projected by individual models in the large ensemble of GCM simulations producing these results before making definitive conclusions about the likelihood of these changes in extremes.

On the daily time scale, an ensemble of simulations from two atmosphere-ocean-coupled GCMs was analyzed; both models simulate a temperature increase on the warmest night of the year that is larger than the mean response over the Amazon basin but smaller than the mean response over parts of southern South America (Hegerl and others 2004). Concerning extreme precipitation, both models foresee a stronger wettest day per year over large parts of southeastern South America and central Amazonia and weaker precipitation extremes over the coasts of northeastern Brazil.

Changes in extremes were analyzed based on multimodel simulations from nine global coupled climate models (Tebaldi and others 2006). Figure A.6 depicts projected changes in extreme precipitation and the number of consecutive dry days (Meehl and others 2007). The general pattern of change suggests that nearly everywhere, precipitation intensity

Figure A.6   Changes in Extremes Based on Multimodel Simulations from Nine Global Coupled Climate Models

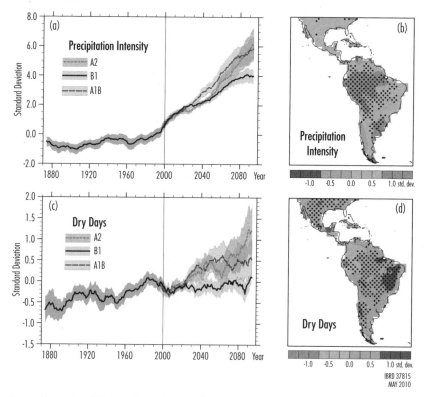

Source: Adapted from IPCC 2007, adapted from Tebaldi and others 2006.
Note: (a) Globally averaged changes in precipitation intensity (defined as the annual total precipitation divided by the number of wet days) for low (SRES B1), middle (SRES A1B), and high (SRES A2) scenarios.
(b) Changes in spatial patterns of simulated precipitation intensity between two 20-year means (2080–99 minus 1980–99) for the A1B scenario.
(c) Globally averaged changes in dry days (defined as the annual maximum number of consecutive dry days).
(d) Changes in spatial patterns of simulated dry days between two 20-year means (2080–99 minus 1980–99) for the A1B scenario.
Solid lines in (a) and (c) are the 10-year smoothed multimodel ensemble means; the envelope indicates the ensemble mean standard deviation. Stippling in (b) and (d) denotes areas where at least five of the nine models concur in determining that the change is statistically significant. Extreme indices are calculated only over land following Frich and others (2002). Each model's time series was centered on its 1980 to 1999 average and normalized (rescaled) by its standard deviation computed (after detrending) over the period 1960 to 2099. The models were then aggregated into an ensemble average, both at the global and at the grid-box level. Thus, changes are given in units of standard deviations.

is increasing, enhancing the risks of flash floods. At the same time, there is a clear tendency toward an increase in the number of consecutive dry days almost everywhere. This suggests an increased risk of droughts. These seemingly opposite tendencies are physically consistent: in a warmer atmosphere, more moisture is available for precipitation in moist air than under present-day conditions. This will lead to an enhanced possibility for more intensive precipitation events. On the other hand, higher temperatures will also make the less-moist air masses even drier, leading to fewer rainy days overall, and will thus increase the chance of longer dry periods.

In Central America and in the Caribbean, a substantial contribution to the intensive precipitation events is connected with tropical cyclone activity. It is therefore relevant to keep in mind how the incidence of such systems may change. Recent studies with improved global models, ranging in resolution from about 100 to 20 km, suggest future changes in the number and intensity of tropical cyclones (hurricanes). A synthesis of the model results to date indicates with a warmer future climate, there will be increased peak wind intensities and increased mean and peak precipitation intensities in future tropical cyclones, with the possibility of fewer relatively weak hurricanes and more numerous intense hurricanes. However, the total number of tropical cyclones globally is projected to decrease. The observed increase in the proportion of very intense hurricanes since 1970 in some regions is in the same direction but is much larger than predicted by theoretical models.

Sea Level

Projected rises in the global-average sea level at the end of the 21st century (2090–99) relative to 1980–99 were estimated by the IPCC for the six SRES marker scenarios. The results, given as 5 to 95 percent ranges based on the spread of model results, are shown in table A.1. Thermal expansion contributes 70 to 75 percent to the best estimate for each scenario. The use of atmosphere-ocean-coupled GCMs to evaluate ocean heat uptake and thermal expansion makes these estimates more credible than previous ones. It has also reduced the projections compared to previous estimates. In all the SRES marker scenarios except B1, the average rate of sea-level rise during the 21st century very likely exceeds the 1961–2003 average rate (1.8 ± 0.5 mm per year). For an average model, the scenario spread in sea-level rise is only 0.02 m by the middle of the century, but by the end of the century it is 0.15 m.

In the IPCC best-estimate assessment, recent claims of changes in accelerating melting from Greenland and ice mass loss from Antarctica

Table A.1   Projected Global Average Surface Warming and Sea-Level Rise at the End of the 21st Century

Case	Temperature change (°C at 2090–99 relative to 1980–99)[a] Best estimate	Likely range	Sea-level rise (m at 2090–99 relative to 1980–99) Model-based range excluding future rapid dynamical changes in ice flow
Constant Year 2000 Concentrations[b]	0.6	0.3 – 0.9	NA
B1 Scenario	1.8	1.1 – 0.9	0.18 – 0.38
A1T Scenario	2.4	1.4 – 3.8	0.20 – 0.45
B2 Scenario	2.4	1.4 – 3.8	0.20 – 0.43
A1B Scenario	2.8	1.7 – 4.4	0.21 – 0.48
A2 Scenario	3.4	2.0 – 5.4	0.23 – 0.51
A1F1 Scenario	4.0	2.4 – 6.4	0.26 – 0.59

Source: IPCC 2007.
Note: (a) These estimates are assessed from a hierarchy of models that encompass a simple climate model, several earth models of intermediate complexity, and a large number of atmosphere-ocean global circulation models (AOGCMs). (b) Year 2000 constant composition is derived from AOGCMs only.

were not considered, largely because of incomplete knowledge. However, values considerably higher than the best estimates—up to a factor two higher by the end of the century—cannot be ruled out. Continuing documentation of the observed changes supports this possibility. Therefore, the IPCC estimates may be somewhat conservative, although there is still a great uncertainty concerning the future changes in both the Greenland and Antarctica ice masses as a consequence of continued rise in temperatures.

Other Time Horizons

To document the proof of concept, most of the climate change and variability projections in IPCC (2007) focused on the late 21st century. Moreover, most attention was given to the A1B SRES marker scenario.[9] To give some qualitative information about the likely changes in time periods closer to the present, that is, the 2020s or the 2050s, it is necessary to introduce scaling arguments, as a full utilization of the available GCM information entering the IPCC assessment work is not practically possible for the present study. As demonstrated by Christensen and others (IPCC 2007), using a scaling approach is reasonable, provided the climate parameters being addressed are robust and vary slowly over time. This further implies that one should address decades rather than a particular year. Thus, investigating

2020 should inform one about the period 2010–30, and estimates for 2050 would refer to the period 2040–60; but even somewhat longer periods would be preferable (box A.1).

Taking a scaling approach would imply that the local amplification factor compared with the global mean temperature change is known. Figure A.5, in combination with the global temperature increase for the A1B scenario, provides the foundation for such a scaling argument.[10] The mean warming between 1990–99 and 2090–99 can be estimated to be 3°C. For the period representing 2020, this number equals 0.5°C, and for 2050, the number is about 1.5°C. These figures correspond to scaling numbers of 0.17 and 0.5, respectively. This can be immediately used to interpret the top row (temperature maps) of figure A.5. Precipitation is somewhat more subtle to deal with; as a gross measure, one may still use the same scaling numbers, keeping in mind that the boundaries between positive and negative change could well change because of nonlinearity in the climate system and natural variability, which tend to dominate as long as the temperature signal is weak.

Considering the projected change in extreme events, such as characterized by the analysis of consecutive dry days and precipitation intensity discussed in connection with figure A., some indications about the scaling can be found from inspecting the time evolution of the globally averaged change. Note that little or no change is found for the 2020s, while by the 2050s the first clear signs of the climate change and variability signal toward the end of the century could be expected, although perhaps only showing through with about 50 percent of the power.

Further Research

To advance the scientific knowledge and awareness of past, contemporary, and future climate for LAC, it is vital that efforts are made to secure or rescue existing databases of climate records in the region. All such data archives, no matter how small, are potentially useful for this purpose. Databases should be made available to the international scientific community—at best free of charge, but otherwise through collaborative international projects to analyze and exploit them.

It is also essential to encourage a stronger involvement of LAC scientists in climate analysis at the international level. At present, too many efforts in climate analysis go only as far as the archives of national meteorological services, although a huge effort has been undertaken by the World Meteorological Organization and the U.S. National Oceanic and

Atmospheric Administration/National Climatic Data Center to push LAC climatologists forward.[11] Only by entering the international science arena will regional findings become useful in assisting mankind to mitigate and adapt to anthropogenic climate change and variability. Regional climate change and variability projections for LAC, based on comprehensive analysis of climate, exist only in the form of output from cause-resolution GCMs, with a few exceptions. There is a strong demand for coordinated research exploring a range of possibilities to provide information on regional or local climate change and variability. This could be achieved by enabling concerted actions with participation from LAC as well as developed nations, taking the approach put forward in such research programs as the North American Regional Climate Change Assessment Program (NARCCAP) for North America (Mearns and others 2005), and the Prediction of Regional Scenarios and Uncertainties for Defining European Climate Change Risks and Effects (PRUDENCE) in Europe, for example (Christensen and others 2007). These initiatives focus on skillful projections of regional climate change and variability, providing not only estimates of changes in mean properties and their variation, but the scientific knowledge that allows for a quantifiable assessment of uncertainties associated with the projections.

Notes

1. Appendix A was prepared by Jens Hesselbjerg Christensen for the companion report, *Reducing Poverty, Protecting Livelihoods, and Building Assets in a Changing Climate: Social Implications of Climate Change in Latin America and the Caribbean* (Verner 2010).

2. A1F1 describes a high-emission scenario with high economic growth, where global population peaks around 9 billion in 2050 and declines to about 7 billion by 2100, with continued high GHG emissions resulting in cumulative emissions from 1990 until 2100 of 2,182 GtC. B1, in contrast, describes a low-emission scenario with a similar development in population figures and also has high economic growth, but where the gains of this economic growth to a large extent are invested in improved efficiency of resource use ("dematerialization"), equity, social institutions, and environmental protection. Even in the absence of explicit interventions to mitigate climate change, the proactive environmental policies lead to relatively low GHG emissions, reaching cumulative levels of 976 GtC between 1990 and 2100. For a more detailed description of the SRES marker scenarios, see Nakicenovic and others (2000).

3. IPCC Data Distribution Center: http://www.ipcc-data.org.

4. The ITCZ is an air mass at the equator receiving direct radiation from the sun and bounded by the two tropics. In the ITCZ, air that ascends because of the heat of the sun is replaced by air from below originating to the north and south of the equator.

5. A jet is a well-confined strong wind that efficiently transfers air masses from one region to another. Other examples of jet streams are found over the Atlantic at high altitudes (10 km), where transatlantic airplanes benefit from their presence when flying eastbound.

6. The contrast between ocean and land masses near the equator results in differential rising and ascending of air masses in a systematic manner. This is referred to as the Walker circulation.

7. Most coupled models continue to simulate ENSO variability during the 21st century. However, slowly the long-term response to warming is a change in the Pacific SST pattern resembling that of El Niño.

8. The A1B scenario, like the other scenarios in the A1 family of scenarios, assumes high economic growth and continued population growth until 2050, when global population peaks around 9 billion people and then declines to about 7 billion people by the end of the 21st century. The A1 family of scenarios also assumes convergence among regions, capacity building, and increased cultural and social interactions, with a substantial reduction in regional differences in per capita income. In terms of changes in energy technology, the A1B scenario assumes balance across all energy sources, meaning not relying too heavily on one particular energy source, on the assumption that similar improvement rates apply to all energy supply and end-use technologies. This places A1B in the midrange with respect to GHG emissions; it predicts carbon dioxide emissions increasing until around 2050 and then decreasing after that. For a more detailed description, see Nakicenovic and others (2000).

9. For a more detailed, description see Nakicenovic and others (2000).

10. Global temperature increase for the A1B scenario as depicted in IPCC (2007a), figure SPM.5 and table TS.6, reproduced above as table A.1

11. See the most recent versions of "State of the Climate" issued in the *Bulletin of the American Meteorological Society*, and Arguez (2007), Levinson (2005), Levinson and Lawrimore (2008), and Shein (2006) in the same journal.

References

Arguez, A., ed. 2007. "State of the Climate in 2006." *Bulletin of the American Meteorological Society* 88: S1–S135.

Bales, R. C., D. M. Liverman, and B. J. Morehouse. 2004. "Integrated Assessment as a Step Toward Reducing Climate Vulnerability in the Southwestern United States." *Bulletin of the American Meteorological Society* 85 (11): 1727.

Berbery, E. H., and E. A. Collini. 2000. "Springtime Precipitation and Water Vapor Flux Convergence over Southeastern South America." *Monthly Weather Review* 128: 1328–1346.

Christensen, J. H., T. R. Carter, M. Rummukainen, and G. Amanatidis. 2007. "Evaluating the Performance and Utility of Regional Climate Models: The PRUDENCE Project." *Climatic Change* 81 (Supl. 1): 1–6, doi:10.1007/s10584-006-9211-6.

Dettinger, M. D., D. S. Battista, G. J. McCabe, R. D. Garreaud, and C. M. Bitz. 2001. "Interhemispheric Effects of Interannual and Decadal ENSO-Like Climate Variation on the Americas." In *Interhemispheric Climate Linkages*, ed. V. Markgraf. San Diego: Academic Press.

Doyle, M. E., and V. R. Barros. 2002. "Midsummer Low-Level Circulation and Precipitation in Subtropical South America and Related Sea Surface Temperature Anomalies in the South Atlantic." *Journal of Climate* 15: 3394–3410.

Emanuel, K. 2003. "Tropical Cyclones." *Annual Review of Earth and Planetary Science* 31: 75–104.

Frich, P., L. V. Alexander, P. Della-Marta, B. Gleason, M. Haylock, A. M. G. Klein Tank, and T. Peterson. 2002. "Observed Coherent Changes in Climate Extremes during the Second Half of the Twentieth Century." *Climate Research* 19: 193–212.

Garreaud, R. D. 2000. "A Multi-Scale Analysis of the Summertime Precipitation over the Central Andes." *Monthly Weather Review* 127: 901–921.

Giorgi, F., and R. Francesco. 2000. "Evaluating Uncertainties in the Prediction of Regional Climate Change." *Geophysical Research Letters* 27: 1295–1298.

Gray, W. M.. 1984. "Atlantic Seasonal Hurricane Frequency. Part I: El Niño and 30 mb Quasi-Biennial Oscillation Influences." *Monthly Weather Review* 112: 1649–1668.

Hegerl, G. C., F. W. Zwiers, P. A. Stott, and V. V. Kharin. 2004. "Detectability of Anthropogenic Changes in Annual Temperature and Precipitation Extremes." *Journal of Climate* 17: 3683–3700.

IPCC (Intergovernmental Panel on Climate Change). 2007. *Climate Change 2007: The Physical Science Basis.* Contribution of Working Group I to the Fourth Assessment Report of the IPCC. Geneva: IPCC. http://www.ipcc.ch.

Kidson, J. W. 1988. "Interannual Variations in the Southern Hemisphere Circulation." *Journal of Climate* 1: 1177–1198.

Levinson, D. H., ed. 2005. "State of the Climate in 2004." *Bulletin of the American Meteorological Society* 86: S1–S86.

Levinson, D. H., and J. H. Lawrimore, eds. 2008. "State of the Climate in 2007." *Bulletin of the American Meteorological Society* 89: S1–S179.

Liebmann, B., G. N. Kiladis, J. A. Marengo, T. Ambrizzi, and J. D. Glick. 1999. "Submonthly Convective Variability over South America and the South Atlantic Convergence Zone." *Journal of Climate* 12: 1877–1891.

Magaña, V., J. A. Amador, and S. Medina. 1999. "The Mid-Summer Drought over Mexico and Central America." *Journal of Climate* 12: 1577–1588.

Mann, M. E., R. S. Bradley, and M. K. Hughes. 2000. "Long-Term Variability in the El Nino Southern Oscillation and Associated Teleconnections." In *El Nino and the Southern Oscillation: Multiscale Variability and Its Impacts on Natural Ecosystems and Society,* eds. H. F. Diaz and V. Markgraf, 321–372. Cambridge, U.K.: Cambridge University Press.

Marengo, J. A., W. R. Soares, C. Saulo, and M. Nicolini. 2004. "Climatology of the Low-Level Jet East of the Andes as Derived from the NCEP-NCAR Reanalyses: Characteristics and Temporal Variability." *Journal of Climate* 17: 2261–2280.

Mearns, L. O., R. W. Arritt, G. Boer, D. Caya, P. Duffy, F. Giorgi, W. J. Gutowski, I. M. Held, R. Jones, R. Laprise, L. R. Leung, J. Pal, J. Roads, L. Sloan, R. Stouffer, E. S. Takle, and W. Washington. 2005. "NARCCAP, North American Regional Climate Change Assessment Program: A Multiple AOGCM and RCM Climate Scenario Project over North America." *Preprints of the American Meteorological Society 16th Conference on Climate Variations and Change, 9–13 January, 2005.* Paper J6.10, 235–238. Washington, DC: American Meteorological Society.

Meehl, G. A., T. F. Stocker, W. D. Collins, P. Friedlingstein, A. T. Gaye, J. M. Gregory, A. Kitoh, R. Knutti, J. M. Murphy, A. Noda, S. C. B. Raper, I. G. Watterson, A. J. Weaver, and Z.-C. Zhao. 2007. "Global Climate Projections." In *Climate Change 2007: The Physical Science Basis,* eds. S. Solomon, D. Qin, M. Manning, Z. Chen, M. Marquis, K. B. Averyt, M. Tignor, and H. L. Miller. Contribution of Working Group I to the Fourth Assessment Report of the Intergovernmental Panel on Climate Change. Cambridge, U.K., and New York: Cambridge University Press.

Mo, K. C., and J. Nogués-Paegle. 2001. "The Pacific-South American Modes and Their Downstream Effects." *International Journal of Climatology* 21: 1211–1229.

Nakicenovic, N., O. Davidson, G. Davis, A. Gr̦bler, T. Kram, E. Lebre La Rovere, B. Metz, T. Morita, W. Pepper, H. Pitcher, A. Sankovski, P. Shukla, R. Swart, R. Watson, and Z. Dadi. 2000. *IPCC Special Report on Emissions Scenarios.* Cambridge, U.K.: Cambridge University Press.

Nogués-Paegle, J., C. R. Mechoso, R. Fu, E. H. Berbery, W. C. Chao, T. C. Chen, K. H. Cook, A. F. Diaz, D. Enfield, R. Ferreira, A. M. Grimm, V. Kousky, B. Liebmann, J. Marengo, K. Mo, J. D. Neelin, J. Paegle, A. W. Robertson, A. Seth, C. S. Vera, and J. Zhou. 2002. "Progress in Pan American CLIVAR Research: Understanding the South American Monsoon." *Meteorologica* 27 (1&2): 1–30.

Robertson, A. W., and C. R. Mechoso. 2000. "Interannual and Interdecadal Variability of the South Atlantic Convergence Zone." *Monthly Weather Review* 128: 2947–2957.

Romero-Centeno, R., J. Zavala-Hidalgo, A. Gallegos, and J. J. O'Brien. 2003. "Isthmus of Tehuantepec Wind Climatology and ENSO Signal." *Journal of Climate* 16: 2628–2639.

Ruosteenoja, K., T. R. Carter, K. Jylha, and H. Tuomenvirta. 2003. *Future Climate in World Regions and Intercomparison of Model-Based Projections for the New IPCC Emissions Scenarios*. Helsinki, Finland: Finnish Environment Institute.

Shein, K. A., ed. 2006. "State of the Climate in 2005." *Bulletin of the American Meteorological Society* 87: S1–S102.

Silvestri, G. E., and C. S. Vera. 2003. "Antarctic Oscillation Signal on Precipitation Anomalies over Southeastern South America." *Geophysical Research Letters* 30 (21): 2115.

Tang, B. H., and J. D. Neelin. 2004. "ENSO Influence on Atlantic Hurricanes via Troposhperic Warming." *Geophysical Research Letters* 31: L24204, doi:10.1029/2004GL021072.

Taylor, M., and E. Alfero. 2005. "Climate of Central America and the Caribbean." In *The Encyclopedia of World Climatology*, ed. J. Oliver. Encyclopedia of Earth Sciences Series. Netherlands: Springer Press. DOI: 10.1007/1-4020-3266-8. ISBN: 978-1-4020-3264-6 (Print) 978-1-4020-3266-0 (Online).

Tebaldi, C., K. Hayhoe, J. M. Arblaster, and G. E. Meehl. 2006. "Going to the Extremes: An Intercomparison of Model-Simulated Historical and Future Changes in Extreme Events." *Climate Change* 79: 185-211.

Thompson, D. W. J., and J. M. Wallace. 2000. "Annular Modes in the Extratropical Circulation. Part I: Month-to-Month Variability." *Journal of Climate* 13: 1000–1016.

Vera, C. S., and P. K. Vigliarolo. 2000. "A Diagnostic Study of Cold-Air Outbreaks over South America." *Monthly Weather Review* 128: 3–24.

Vernekar, A., B. Kirtman, and M. Fennessy. 2003. "Low-Level Jets and Their Effects on the South American Summer Climate as Simulated by the NCEP Eta Model." *Journal of Climate* 16: 297–311.

Verner, Dorte. 2010. *Reducing Poverty, Protecting Livelihoods and Building Assets in a Changing Climate: Social Implications of Climate Change in Latin America and the Caribbean*. Washington, DC: World Bank.

Zhang, X., Francis W. Zwiers, Gabriele C. Hegerl, F. Hugo Lambert, Nathan P. Gillett, Susan Solomon, Peter A. Stott, and Toru Nozawa. 2007. "Detection of Human Influence on Twentieth-Century Precipitation Trends." *Nature* 448: 461–464, doi:10.1038/nature06025.

Field Work Methodology

This section sketches out the fieldwork strategy employed to gather information to understand the social impact of climate change to indigenous peoples in LAC. In the LAC region there are more than 600 different peoples, each with their own language and worldview. They live in and depend on highly diverse environments, within the entire range of political and economic systems of the region. A study like the present one can offer only a partial perspective on the different realities experienced by different indigenous peoples, or on the ways they interpret and respond to climatic changes and variability.

Our overall approach took its departure from a choice of the main types of climatological impact besetting LAC: hurricanes and intensified storms, increasing temperatures with corresponding glacier retreat, increased drought, and unpredictable variations in precipitation regimes. On this basis, we divided LAC into three major ecogeographical regions: the Amazon, the Andes and Sub-Andes, and the Caribbean and Mesoamerica. In each of these regions, we selected areas and peoples of particular interest for further analysis.

We undertook fieldwork in communities and offices in Bolivia, Colombia, Mexico, Nicaragua, and Peru. As most of the changes and types of impact that we analyzed cut across many segments and conditions, we sought to make the information gathering and data analysis responsive to elements such as gender, age, and the social and ecological situation of indigenous peoples, including land availability, degree of contact with national society, and cultural resourcefulness, in order to configure variable impact scenarios.

During fieldwork, the first step, based on the above conceptual framework, involved establishing an overview of relevant socioeconomic, cultural, and natural conditions in the case-study countries through available sources of national and local data. These sources were kindly provided by, or checked with, government officials, researchers from national institutions, and universities. In addition, a number of national and international nongovernmental organizations were consulted. On this basis, a number of communities were selected for further study.

In the next step, local perceptions of the importance of assets and opportunity structures[1] were identified through open-ended interviews with key informants selected through maximum variation sampling. A semistructured questionnaire guided the interviews.

In the third step, scenario analyses with key informants were undertaken. A realistic climate-change impact scenario was established for each region with the help of the Danish Meteorological Institute,[2] based on main characteristics, trends, and projections. This scenario was devised so that it could be related to a specific climate event or climate change tendency that all or most local people could remember and relate to. During the exercise, the scenario was related to specific, relevant aspects and themes in an open-ended fashion that would not provoke biases or "wanting to please the interviewer" situations. Key informants were carefully selected, ensuring that they were representative of different segments of the community based on gender, social standing, and livelihood strategy. The scenario was played out and discussions facilitated and recorded to establish (1) what people did when they experienced the climate change event, (2) which resources (assets) they drew upon, (3) which assets were temporarily or lastingly damaged, (4) which resources (assets) could be used to replace them, and (5) what people thought would have improved their ability to cope and adapt during that climate event. The interviewer asked about the institutions and networks that some individuals and families were able to draw upon to increase the effect of their coping strategies, and related them to the institutional mapping.

In the final step, institutional mapping was carried out. A Venn institutional analysis (FAO and SEAGA, 1998)[3] was used to supplement the information from the key informant interviews and scenario analysis described above. The aim was to understand in more depth the assets and opportunity structures upon which people's coping and adaptation strategies depended. The exercise involved drawing a circle and describing the center as the most important place for help to adapt to changes, with the importance declining with movement away from the center. Then the following three types of questions were posed (depending on the weather-related event described): (1) "Which institutions were present during this event? Please describe them briefly." (2) "Which was the most important institution, in your opinion, for supplying information, relief, and rehabilitation"? (3) "Please order the rest of the institutions according to their importance for you and the community." Finally, a secondary information review was continuously undertaken based on a gap analysis.

Notes

1. Both assets and opportunity structures are fundamental dimensions when analyzing types and levels of social impact.

2. Specifically, Jens Hesselbjerg Christensen.

3. See also www.idrc.ca.

Reference

FAO (Food and Agricultural Organization) and SEAGA. 1998. "Venn Diagrams; Institutional Profiles." In *Field Handbook: Socioeconomic and Gender Analysis Program*, Tools A5 and A6. Rome: FAO. http://www.fao.org.

Index

Boxes, figures, notes, and tables are indicated by b, f, n, or t following the page number.